Food Contact Materials – Rubbers, Silicones, Coatings and Inks

Martin Forrest

*i*Smithers – A Smithers Group Company

Shawbury, Shrewsbury, Shropshire, SY4 4NR, United Kingdom
Telephone: +44 (0)1939 250383 Fax: +44 (0)1939 251118
http://www.ismithers.net

First Published in 2009 by

*i*Smithers

Shawbury, Shrewsbury, Shropshire, SY4 4NR, UK

©2009, Smithers Rapra

A catalogue record for this book is available from the British Library.

Every effort has been made to contact copyright holders of any material reproduced
within the text and the authors and publishers apologise if
any have been overlooked.

ISBN (Hard-backed): 978-1-84735-141-8
ISBN (Soft-backed): 978-1-84735-414-3

Typeset by *i*Smithers
Indexed by Damco Solutions, Hayes, Middlesex, UK
Printed and bound by Lightning Source

Contents

Food Contact Rubbers – Products, Migration and Regulation

Coatings and Inks for Food Contact Materials

Silicone Products for Food Contact Applications

Preface

For many years, Smithers Rapra Technology has carried out research projects for the UK Food Standards Agency (FSA) and its precursor, The Ministry of Agriculture, Fisheries and Food (MAFF). These have covered a wide range of polymer products (e.g., rubbers, silicone based materials, ion-exchange resins, laminate materials etc) and have provided UK Government Agencies with important information on the materials and the manufacturing practices that are used in industry, as well as making an important contribution to the data, via extensive scientific investigations, that is available with respect to the migratory behaviour of these products when they are in contact with food.

This book is an amalgamation of three recent Rapra Review Reports - numbers 182, 186 and 188. Each of these original reports had as its origin one or more of the Rapra managed MAFF/FSA projects referred to above. In addition to providing the wealth of information that is described in the paragraph below, this publication has therefore effectively brought together the results and findings of the following research projects:

a) Migration data on Food Contact Rubbers (1994 to 1997) – Contract FS2219

b) Further Migration Data on Food Contact Rubbers (1997 to 2001) – Contract FS2248

c) Rubber Breakdown Products (2002 to 2004) – Contract A03039

d) Chemical Migration from Silicones used in connection with Food Contact Materials and Articles (2003 to 2005) – Contract A03046

e) An Assessment of the Potential of Migration of Substances from Inks and their respective Coatings (2005 to 2007) – Contract A03055

Although each of the reviews takes into account the unique nature of its titular food contact product grouping, all three have been written with a similar framework and structure in mind to facilitate the cross-referencing of important topics and technical areas, e.g., EU legislation and migration testing regimes and methodologies.

The overall objective of each review is to provide an expert overview of the products as food contact materials, with a comprehensive accompanying list of relevant references from the Smithers Rapra Polymer Library to enable further reading. All of the salient areas of the subject have been addressed within the review section. In each case there is an initial in-depth description of the variety and types of products that are used in industry, and the chemical processes associated with their manufacture. The behaviour of these materials in service is closely related to their chemical and physical properties, and so an appropriate level of attention is given to these subjects, and the complexities and techniques of product formulation. A summary of the relevant food contact regulations, both within the European Economic Area (e.g., EU and CoE) and the USA, together with the migration and analytical testing regimes used to assess their suitability for food contact are included to assist end-users and food scientists in their decision making processes. To this end, an appraisal of the potential migrants (and their origins) that are present in the materials is also provided, together with a concise summary of the migration data that has been published. In addition, to aid workers in research and development roles, recent technological developments that have taken place are described, particular those aimed at improving food safety, as are the overall trends that have been identified and their expected future direction and commercial impact.

With the exception of new, concise introductory chapters to take into account this new format, each review is reproduced here in its unabridged, original form. This publication therefore brings together these three important sources of food contact information in a single, convenient volume and will be an important reference source for workers in the food industry in general, and within the food contact field in particular.

Dr Martin Forrest

Smithers Rapra Technology, Shawbury, May 2009

Introduction

Food Contact Rubbers – Products, Migration and Regulation

This review is a completely revised and updated version of a review (Rapra Review Report number 119), which was published in 2000. Since that time a number of important developments have taken place, notably the beginning of the harmonisation of the legislation within Europe with the issuing of the CoE Resolution on rubber in 2004. Over the years, the FSA has commissioned three fundamental studies at Rapra into the potential migrants that are present in rubber products (contracts FS2219, FS2248 and A03038), with silicone rubbers being included within the scope of contract A03046. The results from all four of these projects have been included within this review.

A rubber can be defined as a polymeric material which, above its glass transition temperature, can be stretched repeatedly to at least twice its original length and, on release of the stress, rapidly returns to its original length. These properties are brought about by a combination of the chemical structure of the polymer backbone and the vulcanisation process which brings about the formation of a lightly cross linked three dimensional structure. Thermoplastic rubbers also fall within the scope of this review report. These are a group of polymers that are similar to rubbers in resilience and rapid recovery, but can be repeatedly softened by heating (i.e., are thermoplastic) to enable processing, and regain their elastomeric character on cooling to room temperature. There is also a sub-set of materials named thermoplastic vulcanisates (TPV) in which the rubber phase within the material is cross linked, but this is done at the polymerisation stage, and these materials do not undergo a vulcanisation step.

In contrast to plastics, rubbers are rarely used in the packaging of food products. Examples where there are found include the use of rubber in flip top seals on beer bottles, rubber seals used in some jar tops (e.g., paste containers) and the seal that is present in food cans. However, in the processing of food, there are a number of situations where significant contact of the food with rubber products can occur. This is due to the fact that the unique properties of rubber lead to it being used in a wide range of products including conveyor belting, hosing, seals, gaskets, skirting and specific products such as milk liners (see Section 2.3). It is also the case that the range of contact conditions encountered (i.e., food type, contact temperature, time and area) mean that a wide variety of rubber types are employed (Section 2.1). The contact times with food in processing situations are short (e.g., minutes) and the contact areas, apart from hose and belting, are small. This is in contrast to plastics which, when used as packaging materials, often have long contact times (e.g., weeks) and relatively large contact areas.

It is also the case that their inherent properties and processing requirements mean that rubber compounds are normally more complex than plastics. For example, it is common practice for a rubber formulation to contain additives such as process aids and plasticisers, antidegradants, a curative, and cure co-agents and accelerators, resulting in a list of from ten to fifteen ingredients. In order to achieve the desired final properties, it is also relatively common for the base polymers to be blended, particularly the diene (e.g., natural rubber, SBR, polybutadiene) type rubbers. The consequence of these considerations is that, compared to plastics, there will be a larger range of monomers, oligomers and polymer related substances (e.g., polymerisation catalysts and process aids) having the potential to migrate into food. Another very important consideration with rubbers is that chemical reactions take place during the curing process and further chemical modification of the matrix occurs due to the action of antidegradants. Both of these processes result in the generation of low molecular weight reaction products and breakdown products. The combined effect of all of these factors is to greatly complicate the process of predicting what has the potential to migrate from a rubber product into food. It is this need for knowledge in these areas that has led the UK's Food Standard Agency (FSA) to fund a number of research projects at independent research organisations such as Rapra to look into the use of rubber as a food contact material.

Silicone Products for Food Contact Applications

The objective of the FSA Silicone food contact materials project (Contract A03046) was to provide detailed information on the types and composition of silicone based products that are used in contact with food and identify the extent to which the migration of specific constituents into food could occur. It built on information previously obtained on silicone food contact materials, such as seals and tubing, by routes such as the MAFF Rubber contact rubber project FS2219.

Silicones are used in a variety of different food contact situations and conditions. The silicone class of polymer is very versatile and the physical form of a silicone product can vary from relatively low molecular weight lubricants and oils, through high molecular weight rubbery polymers to extensively crosslinked hard resins.

At the time of writing, there is no specific harmonised legislation for food contact silicone materials. However, they are covered by Regulation (EC) No 1935/2004 and a Council of Europe Resolution on Silicones (Resolution AP 2004) but this latter document is only intended to provide guidance to industry and is not legally binding.

As the breadth and scope of this review is greater than the FSA project and, because of the limitations of this format, it has only been possible to summarise the extensive amount of information that was obtained during the course of it. The reader is therefore recommended to apply to the FSA Library at 125 Kingsway, London for a full version of the final report, which was published in January 2005.

Coatings and Inks for Food Contact Materials

The objective of FSA Coatings and Inks project (Contract A03055) was to assess the potential for the migration of substances from coatings and inks that were used in food packaging applications. As a significant amount of work had already been carried out on coatings that were in direct contact with food (e.g., can coatings), a boundary that was set was that only coatings and inks in non-direct food contact situations would be considered. As the scope of this review is greater than the Rapra project and, due to the limitations of this particular format, it has only been possible to include some of the information that was acquired during the course of the FSA project. If the reader has a particular interest in coatings and inks used in these types of applications, they are therefore recommended to apply to the FSA for a full version of the final project report, which was published in March 2007. The contact details for the FSA are given in Section 11.

Coatings and Inks for use with food have been a very topical subject over the last couple of years, mainly due to the culmination of the work that has been carried out by the Council of Europe (CoE). As a result of its efforts, we have seen the adoption of both a Resolution for Coatings, and a Resolution for Inks used on non-food contact surfaces. The Inks Resolution has been controversial with industry bodies throughout Europe, who have claimed that its Inventory list is incomplete and not representative of current industry practice (see Section 5.2). In addition to these regulatory developments, this is an active area for research, with a number of innovative and sophisticated products finding commercial applications, e.g., in active and intelligent packaging, and antimicrobials – see Section 9.

This review has attempted to cover all of the coatings and inks products used in food contact scenarios. Hence, direct and non-direct contact situations are included throughout the food chain, e.g., harvesting, processing, transportation, packaging and cooking. In practice, this encompasses an extremely wide range of polymer systems and formulations, and an emphasis has been placed on coatings and inks used in food packaging, as this is usually regarded as representing the most important application category with respect to the potential for migration to occur, All three of the major material classes used in food packaging products, i.e., metal, paper and board, and plastic, are covered by this review.

Food Contact Rubbers – Products, Migration and Regulation

Food Contact Rubbers – Products, Migration and Regulation

1 Introduction

This review is a completely revised and updated version of a review [1], which was published in 2000. Since that time a number of important developments have taken place, notably the beginning of the harmonisation of the legislation within Europe with the issuing of the CoE Resolution on rubber in 2004. Over the years, the FSA has commissioned three fundamental studies at Rapra into the potential migrants that are present in rubber products (contracts FS2219, FS2248 and A03038), with silicone rubbers being included within the scope of contract A03046. The results from all four of these projects have been included within this review.

A rubber can be defined as a polymeric material which, above its glass transition temperature, can be stretched repeatedly to at least twice its original length and, on release of the stress, rapidly returns to its original length. These properties are brought about by a combination of the chemical structure of the polymer backbone and the vulcanisation process which brings about the formation of a lightly cross linked three dimensional structure. Thermoplastic rubbers also fall within the scope of this review report. These are a group of polymers that are similar to rubbers in resilience and rapid recovery, but can be repeatedly softened by heating (i.e., are thermoplastic) to enable processing, and regain their elastomeric character on cooling to room temperature. There is also a sub-set of materials named thermoplastic vulcanisates (TPV) in which the rubber phase within the material is crosslinked, but this is done at the polymerisation stage, and these materials do not undergo a vulcanisation step.

In contrast to plastics, rubbers are rarely used in the packaging of food products. Examples where there are found include the use of rubber in flip top seals on beer bottles, rubber seals used in some jar tops (e.g., paste containers) and the seal that is present in food cans. However, in the processing of food, there are a number of situations where significant contact of the food with rubber products can occur. This is due to the fact that the unique properties of rubber lead to it being used in a wide range of products including conveyor belting, hosing, seals, gaskets, skirting and specific products such as milk liners (see Section 2.3). It is also the case that the range of contact conditions encountered (i.e., food type, contact temperature, time and area) mean that a wide variety of rubber types are employed (Section 2.1). The contact times with food in processing situations are short (e.g., minutes) and the contact areas, apart from hose and belting, are small. This is in contrast to plastics which, when used as packaging materials, often have long contact times (e.g., weeks) and relatively large contact areas.

It is also the case that their inherent properties and processing requirements mean that rubber compounds are normally more complex than plastics. For example, it is common practice for a rubber formulation to contain additives such as process aids and plasticisers, antidegradants, a curative, and cure co-agents and accelerators, resulting in a list of from ten to fifteen ingredients. In order to achieve the desired final properties, it is also relatively common for the base polymers to be blended, particularly the diene (e.g., natural rubber, Styrene-butadiene rubber (SBR), polybutadiene) type rubbers. The consequence of these considerations is that, compared to plastics, there will be a larger range of monomers, oligomers and polymer related substances (e.g., polymerisation catalysts and process aids) having the potential to migrate into food. Another very important consideration with rubbers is that chemical reactions take place during the curing process and further chemical modification of the matrix occurs due to the action of antidegradants. Both of these processes result in the generation of low molecular weight reaction products and breakdown products. The combined effect of all of these factors is to greatly complicate the process of predicting what has the potential to migrate from a rubber product into food. It is this need for knowledge in these areas that has led the UK's Food Standard Agency (FSA) to fund a number of research projects at independent research organisations such as Rapra to look into the use of rubber as a food contact material.

2 Rubber Materials and Products used in Contact with Food

2.1 Polymers Used in Food Contact Rubbers

As part of the FSA project FS2219 an extensive review of the food industry was carried out by consultants at Rapra Technology Ltd in order to obtain information on the types of rubber used in the food industry. All the major sectors of the food industry were surveyed. Some examples are given next:

Dairy products
Brewery and soft drinks

Abattoir and meat processors

Vending machine dispensers

Food canning and the preserves industry

Food packaging

A relatively large range of rubber types were found to be used in a wide variety of rubber products (see Section 2.3). One of the most important criteria that is used in selecting a particular rubber for a specific end-use application is its temperature resistance, although other properties such as chemical inertness, the physical properties of the resulting product (e.g., tensile strength, abrasion and tear resistance) and the types of additives that have to be incorporated into it to achieve the desired processing and final properties are also considered. The most important classes of rubber used in the food industry are reviewed below.

2.1.1 Natural Rubber (i.e., cis-1,4-polyisoprene)

Food contact natural rubber product compounds are mainly used for gloves, can sealants, teats and soothers, although it is coming under pressure from synthetic polyisoprene and other rubbers (e.g., thermoplastic rubbers and silicones) due to the increasing incidence of protein allergies. In food processing equipment natural rubber products will be found in belting and hosing products, sometimes in blends with other rubbers such as SBR. These rubbers are typically used with aqueous foods under flow or short-term static conditions at low temperatures (<40 °C). The maximum temperature limit for the prolonged use of these products is around 80 °C.

2.1.2 Nitrile Rubber

Nitrile rubber is widely used in compounds designed for seals and gaskets, and in hoses for both aqueous and fatty foods. In particular, dairy hosing and milk liners are normally manufactured in nitrile rubber or nitrile rubber blends (e.g., with SBR). Nitrile rubber is better able to withstand heat ageing than natural rubber and so the maximum continuous use temperature is higher at 120 °C. In practice most applications involve flow or short term static conditions at temperatures below 40 °C. Examples of nitrile rubber compounds that are used with food are provided by De Coster and Magg [2].

2.1.3 Ethylene-propylene Rubber

The principal use of ethylene-propylene rubbers (ethylene-propylene-diene terpolymer

(EPDM) or ethylene-propylene (EPM) types) is in the manufacture of heat exchanger gaskets. When cured using peroxides, these materials can be used for extended periods at up to 150 °C. Normal conditions of service are high temperatures (<130 °C) and flow or static exposure to aqueous food products (e.g., beer).

2.1.4 Fluorocarbon Rubber

There are a number of sub-grades of fluorocarbon rubber (copolymers, terpolymers and tetrapolymers) and they are used in applications where the temperatures would degrade EPM rubber products. They are able to withstand prolonged use at temperatures up to 200 °C. Typical conditions are high temperature (<150 °C) gaskets under flow or static conditions, in contact with aqueous or fatty foods (including oils).

2.1.5 Silicone Rubber

Most of the silicone rubbers used in the food industry are based on polydimethyl vinyl silicone and these materials have very good high and low temperature properties. It is their high temperature resistance that enables them to be used for seals and tubing, for example in drinks vending machines up to 100 °C. Cold cured silicones are used as release coatings on items such as food transportation belts and for sweet moulds.

2.1.6 Thermoplastic Elastomers

These materials, of which there are a number of polymer types, are crosslinked at room temperature, usually due to the influence of 'physical crosslinks' formed by part of the matrix being below its glass transition or crystalline melting temperature, but become thermoplastic at processing temperatures (e.g., >150 °C) and can therefore be processed in the same way as plastics. The fact that they are not therefore thermoset materials affects their working temperature range and restricts it to less than 70 °C. Thermoplastic elastomers (TPE) are used in a variety of food contact products, for example flexible lids [(styrenics, e.g., SBS or styrene-isoprene-styrene (SIS)], and belting, gaskets and tubing (particularly polyurethane types). Other types of TPE that are available include olefinic blends of polypropylene and ethylene-propylene rubber, polyamides and polyesters.

2.1.7 Other Types of Rubbers

In addition to the main groups of rubbers covered above there are also a number of other types that are used in the food industry.

These include:

Butyl rubber – used for articles such as stoppers and seals in contact with aqueous foods;

Polychloroprene rubber – used in articles such as conveyor belts for food transportation;

Acrylic and Hydrin rubbers – specialty materials chosen when the food/contact conditions combination requires their specific properties (e.g., chemical inertness combined with relatively good heat stability).

2.2 Additives Used in Food Contact Rubbers

Rubber technology is a mature science with a history going back some 150 years or more. Over the years a number of scientific discoveries (e.g., vulcanisation with sulfur to increase resilience and recovery and the use of antioxidants to lengthen service life) have contributed to the material's dominance in applications where elasticity/recovery upon deformation combined with durability are required.

This section deals with the major additive groupings that are added to food contact rubber compounds in order to ensure that they possesses the correct properties to be processed, have the physical properties appropriate for the application, and ensure that they have sufficient stability and resistance to ageing in service.

2.2.1 Plasticisers/Process Oils and Fillers

Plasticisers and process oils are used to reduce the viscosity of the compound to aid processing and to modify the physical properties of the final product, e.g., reduce hardness and tensile strength, and increase elongation. In common with the other major classes of additive, there is a wide range of substances to choose from, e.g., phthalates, adipates, sebacates, sulfonates and hydrocarbon oils.

Fillers can be regarded in many ways as having the opposite effect to plasticisers; they increase compound viscosity and hardness and reduce elongation. The principal filler for rubber is carbon black as it interacts with the rubber molecules through rubber-filler interactions which produce a large increase in strength. Other fillers such as silica and the silicates also improve properties, but a number (e.g., calcium carbonate) are mainly used to just adjust hardness and reduce cost.

2.2.2 Curatives and Antidegradants

There are two main classes of curative used in rubber – elemental sulfur and peroxides. The first of these is by far the most important and has a complete technology (cure accelerators and co-agents - see below) built up around it. Other types of curative are used in rubber, e.g., sulfur donors, amines, metal oxides. All these types can be used with food contact rubbers.

A number of compounds (called accelerators) are used to modify the chemistry of a curing reaction to ensure that a rubber achieves a good state of cure in a reasonable time at a convenient temperature. The greatest number of accelerators are associated with sulfur cures, with many different chemical types used, e.g., guanidines, sulfenamides and thiazoles.

It is often the case that a blend of accelerators (slow and fast types) is used to optimise the curing reaction by controlling properties such as induction time, cure rate and extent of cure. This practice has implications for food contact use as reactions between these accelerators and the curative/coagent, and between themselves, can produce low molecular weight species with the potential to migrate into food. Typical cure co-agents include zinc oxide and stearic acid for sulfur systems, and trially cyanurate for peroxides. Coagents differ from accelerators in that their main function is to improve the efficiency of the curing reaction, so that lower amounts of curative can be added, which improves final properties such as ageing resistance, and not to increase the rate of cure.

With respect to antidegradants, these are two main classes of antidegradant used in rubber – antioxidants and antiozonants. As most rubbers contain carbon black filler, which acts as an ultraviolet light (UV) screen, the use of UV stabilisers is extremely rare. There are a number of antidegradants available, the two main classes being amines (staining) and phenolics (non-staining), and a mixture of two or more is often used to confer maximum protection. Other types such as thioesters (mainly used to stabilise the base rubber), phosphates and micro-crystalline waxes can also be used. As the antidegradants protect the rubber in service, reaction/breakdown products result and these, in addition to the products formed from the cure system species, have been the subject of a recent FSA research project (FSA contract number A03038).

2.2.3 Miscellaneous Additives

Other additives which can be used include pre-vulcanisation inhibitors (to reduce the possibility of cure occurring during the mixing and forming stages), coupling agents (to promote filler-rubber interactions, deactivators (e.g., polyethylene glycol) to stop accelerators becoming adsorbed onto the surface of polar fillers such as silica, and bonding agents to assist with fabric to rubber interactions in composite products.

Formulations for a number of typical food contact rubber compounds are given in Sections 4.5.1 and 4.5.2.

2.3 Rubber Products Used in the Food Industry and the Contact Conditions

The majority of the information presented in this section was obtained during the comprehensive survey of the food industry that was undertaken during the course of the FSA research project FS2219. The results obtained during this section of the project are summarised here [1].

2.3.1 Types of Rubber Product

The principal types of rubber products that are used in contact with food are shown in **Table 1**.

Table 1. Examples of rubber components used in food contact applications	
Location	Component(s)
Food transportation	Conveyor belts, hoses, rubber skirting and rubber paddle lips
Pipe work components	Seals, gaskets, flexible connectors and butterfly valves
Pumps	Progressive cavity pump stators, diaphragm pumps
Plate heat exchangers	Gaskets
Machinery/storage vessels	General seals and gaskets
Cans/bottles	Bottle seals and can seals
Food handling/preparation	Gloves and feather pluckers
Food manufacturing	Silicone sweet moulds, rubber squeeze rollers
Food wrapping	Meat and poultry nets

2.3.2 Contact Areas

Food contact areas of whole assemblies, be these total pipelines, paddle lips, or plate heat exchangers cover a wide range of areas from less than 100 cm^2 to around 56,000 cm^2. The highest contact areas were found generally to be associated with plate heat exchangers. The majority of individual rubber components have food contact areas of less than 1000 cm^2, with around two-thirds of these having contact areas of less than 100 cm^2. Individual rubber components usually having contact areas of less than 200 cm^2 include general seals and gaskets, plate heat exchanger gaskets, pipe work seals and packaging seals. Pump components, pipe valves and flexible connectors have greater individual contact areas, up to 1000 cm^2. Individual components that have contact areas greater than 1,000 cm^2 tend to have a high aspect ratio, e.g., hoses and conveyor belts. The highest contact area found in the survey was for a hose at approximately 50,000 cm^2. However, certain dry food conveyor belts (e.g., for barley in a maltings) had potential contact areas of up to 1,500,000 cm^2.

2.3.3 Contact Times

Contact times with individual rubber components tend to be low (i.e., less than 60 seconds) and even in assemblies where a number of components are situated the total contact times are still relatively short, e.g., no more than 3 minutes in a heat exchanger assembly. Few components or assemblies give longer contact times, the exceptions being conveyor belts and sweet moulds where contact times were up to one hour, butterfly valves sealing off soaking tanks (up to 24 hours), beer engine seals (up to 12 hours), and beer ke.g. seals (up to 12 weeks). Meat and poultry netting can remain in contact with the product for up to four weeks. The longest potential contact times of up to five years are associated with certain types of packaging seals, particularly can seals.

2.3.4 Contact Temperatures

The temperature at which food products contact rubber components rarely exceeds 80 °C. Temperatures in the range 100 to 140 °C do occur in some processes, e.g., pasteurisation of UHT milk, the production of toffee in silicone moulds and the sterilisation of cans and jars, but the contact time of the food at the elevated temperatures is normally reasonably short (< one hour). Higher temperatures than these are encountered in the refining of vegetable oils, where temperatures in the range 170-250 °C are used in the deodorising section of the plant, and meat and poultry nets may be subjected to temperatures of up to 250 °C for several hours if they are not removed prior to cooking.

An overall summary of the data collected during the survey of the food industry is shown in **Table 2**

Table 2. Summary of the industry survey on rubber in contact with food

Component	Contact area – individual components (cm³)	Contact area – component assemblies (cm³)	Typical contact times	Maximum contact time	Contact temperature (°C)	
					General	Extreme
Food[1] transportation	Up to 50,000	-	<1 hr	-	<85	-
Pipework components	<1,000	<10,000	<1 hr	2 weeks[2]	<140	250[3]
Pumps	<10,000	-	<1 hr	2 weeks[2]	<85	-
Plate heat exchanger gaskets	<1,000	Up to 56,000	<3 min	-	<140	250[3]
General seals and Gaskets	<1,000	Up to 30,000	<1 hr	12 weeks[2]	<85	-
Packaging seals and closures	<1,000	-	Up to 5 years	-	ambient	140[4]
Miscellaneous	<1,000	-	<1 hr	4 weeks[5]	<85	250[6]

[1] Potential contact areas could be up to 1,500,000 cm³ for some of the longer conveyor belts
[2] May have extended contact times during shut downs. Beer keg seals may have contact times of up to 12 weeks
[3] Refining of vegetable oil
[4] During pasteurisation/sterilisation
[5] Meat and poultry nets
[6] Meat and poultry nets if not removed before cooking

3 Regulations Covering the Use of Rubber as a Food Contact Material

3.1 European Union Legislation

At the moment there is no specific EU legislation for rubber food contact materials or articles (other than nitrosamines in babies' dummies [3]). Thus all such materials or articles need to comply with the general Framework Directive 89/109/EEC [4, 5, 6] so that in normal use they will not transfer their constituents to food in quantities that could endanger health or cause unacceptable changes in the composition of food or deterioration in its organoleptic properties (i.e., taste, texture, aroma, or appearance).

3.2 Council of Europe (CoE) Resolution on Rubber Products

The CoE (an international organisation, separate from the European Union) has a committee of Experts on Materials and Articles Coming into Contact with Food, that meets under the auspices of the Partial Agreement in the Social and Public Health Field. Once adopted, the resolutions and supporting documents drawn up by these bodies are not legally binding, but members of the Partial Agreement, such as the UK, are expected to take note of them.

The CoE Rubber Resolution on food contact elastomers (APRes 2004) was adopted in 2004. This document has an inventory list of additives within it and a small section that deals with breakdown products – nitrosamines and amines. The inventory list is described as Technical document No. 1 – List of substances to be used in the manufacture of rubber products intended to come into contact with foodstuffs. An FSA project (Contract number A03038 – see Section 4.5.3) was commissioned to study the breakdown and reaction products from the curatives and antidegradants present in this inventory list.

3.2.1 Technical Documents

In addition to Technical Document No. 1, there are four other documents associated with the Resolution; they are:

- Technical document No. 2: Guidelines concerning the manufacture of rubber products intended to come into contact with foodstuffs

- Technical document No. 3: Good manufacturing practices of rubber products intended to come into contact with foodstuffs

- Technical document No. 4: Test conditions and methods of analysis for rubber products intended to come into contact with foodstuffs

- Technical document No. 5: Practical guide for users of Resolution APRes (2004) on rubber products intended to come into contact with foodstuffs

3.2.2 Product Categories

The Resolution places rubber products into one of three categories:

Category I comprises the following rubber products for which migration testing is required:

- feeding teats

- rubber products to come in to contact with baby food, for which the R-total is equal or greater than 0.001 (a definition of R-total is given below).

Category II comprises rubber products for which the R-total is equal or greater than 0.001 and for which migration testing is required.

Category III comprises rubber products for which R-total is smaller than 0.001 and for which migration testing is not required, except for rubber products containing nitrosamines, nitrosatable substances or aromatic amines and Category III substances with substances with a specific migration limit (SML) in Technical document No. 1.

These three categories take into account the wide variety of applications for which rubber products are used and the fact that migration may vary with the application. The level of migration for rubber products may be estimated by taking into account four factors, R_1, R_2, R_3 and R_4 referring respectively to the relative contact area, contact temperature, contact time and number of times that the article is used. Categories are based on the intended use or on the result of the multiplication of the four factors ($R_1 \times R_2 \times R_3 \times R_4$ = R total).

3.2.3 R Factors

The factors R_1, R_2, R_3 and R_4 can be defined as follows:

R_1 refers to the relative contact area (A_R) between rubber products and food or beverage, expressed in cm^2 of rubber surface per kg of food or beverage. For a relative area smaller or equal to 100 cm^2/kg foodstuffs, R_1 has a value calculated according to the formula: $R_1=A_R/100$. For a relative surface larger than 100 cm^2/kg, R_1 always has the value of 1.00.

R_2 refers to the temperature during the contact period of the rubber product with the food or beverage. At a temperature lower than or equal to 130 °C, R_2 has a value calculated according to the formula: $R_2 = 0.05e^{0.023T}$. Where 'e' is the base of the natural or Napierian logarithms and T is the contact temperature, expressed in °C. For temperatures higher than 130 °C, R_2 always has the value 1.00.

R_3 refers to the time, t, expressed in hours, during which a rubber product is in contact with the food or beverage. For a contact time shorter than or equal to 10 hours, R_3 has a value calculated according to the formula: $R_3 = t/10$. For a contact time of more than 10 hours, R_3 has the value 1.00.

R_4 refers to the number of times, N, that one and the same rubber product, or part of that rubber product comes into recurrent contact with a quantity of food or beverage. If the number of contact times is greater than 1000, then R_4 is calculated according to the formula: $^{10}\log R_4 = 6 - 2 \ ^{10}\log N$. If the number of contact times is smaller than or equal to 1000, then R_4 always has the value 1.00.

The Resolution also states that, amongst other things:

1. Rubber products in Categories I and II should not transfer their constituents to foodstuffs or food simulants in total quantities >60 mg/kg food or food simulant (overall migration limit).

2. Rubber products of Categories I and II should comply with the restrictions laid down in Technical document No. 1. In addition these rubber products should comply with the requirements set out in Table 1 of the Resolution, excepting rubber teats which should comply with Directive 93/11/EEC.

3. Rubber products intended for repeated use should be subjected to tests according to Directive 2002/72/EC Annex 1 [7].

4. Rubber products belonging to Category III do not require migration testing, unless otherwise specified.

5. Verification of compliance with the quantitative restrictions should be carried out according to the requirements laid down in Technical document No. 2 – Practical guide for users of Resolution AP (2004) on rubber products intended to come into contact with foodstuffs. This is an accompanying document to the Resolution and provides more detail regarding the following:

   • further definitions and data surrounding the R values

   • examples of rubber products that fall into Categories II and III and the calculations applied to them

   • migration tests

3.2.4 Silicone Rubbers

There is a separate CoE Resolution (APRes 2004) covering silicone materials for food contact. The resolution defines the silicone product group being comprised of silicone rubbers, silicone liquids, silicone pastes and silicone resins. Blends of silicone rubber with organic polymers are covered by the resolution where the silicone monomer units are the predominate species by weight. Silicones that are used as food additives (e.g., as defoamers in the manufacture of substances such as wine) are not covered by this resolution, but polysiloxanes used as emulsifiers are. The resolution gives an overall migration limit of 10 mg/dm^2 of the surface area of the product or material, or 60 mg/kg of food. There are restrictions on the types of monomers that can be used to produce the silicone polymers and there is an inventory list - Technical document No. 1 - *List of substances used in the manufacture of silicone used for food contact applications.*

3.3 Food and Drug Administration (FDA) in the USA

In America the FDA produces a Guidance for Industry document entitled '*Preparation of Food Contact Notifications and Food Additive Petitions for Food Contact Substances: Chemistry Recommendations*'. This is in addition to the *Code of Federal Regulations Volume 21, Parts 170 to 199 Food and Drugs*, which contains the FDA food contact regulations. This is published annually and rubber products for use with food are covered in Part 170, specifically *Rubber articles intended for repeat use*: 177.2600; and *Closures with sealing gaskets for food containers*: 177.1210. A review and comparison of the FDA and German regulations (Section 3.4) with particular emphasis on the inventory lists of approved ingredients has been carried out by Pysklo [8]. A schedule of commercial ingredients which meet the requirements of the two regulations is provided.

The FDA regulations are relatively straightforward. Providing that the ingredients in the rubber are listed as being approved, and the water (for aqueous food use) or hexane (for fatty food use) extractables under reflux conditions are within the prescribed limits, then the compound is considered suitable for food use (see **Table 3**).

In addition to listed compounding ingredients, the regulations also allow the use of prior sanctioned ingredients and also additives that are generally recognised as safe (GRAS). Prior sanctioned materials are listed in the appropriate sections of the FDA regulations, for example:

181.27	Antioxidants	181.28	Release agents
181.27	Plasticisers	181.29	Stabilisers

Table 3. FDA migration limits for repeat use articles	
Fatty foods – Hexane extractables under reflux	
First seven hours	175 mg/in^2
Succeeding two hours	4 mg/in^2
Aqueous foods – Distilled water extractables under reflux	
First seven hours	20 mg/in^2
Succeeding two hours	1 mg/in^2

Other indirect food substances described as GRAS, which can be used as rubber additives, are listed in various parts of CFR 21. Substances listed in section 182 include zinc oxide (182.8991), zinc stearate (182.8994) and calcium silicate (182.2227). GRAS food substances listed in section 184 include calcium carbonate, calcium stearate and calcium oxide. Other multi-purpose GRAS food substances include kaolin clay (186.1256) and iron oxide (186.1374).

The FDA regulations specifically prohibit the use of the following ingredients in food contact rubbers:

Section 189.220   Polymerised 1,2-dihydro-2,2,4-trimethylquinoline

Section 189.250   Mercaptoimidazoline and 2-mercaptoimidazoline

The FDA also places severe restrictions (i.e., 0.003 /in^2) on the migration of acrylonitrile monomer from nitrile rubbers (Section 181.32).

3.4 Bundesinstitut für Risikobewertung (BfR) German Regulations

Within Germany, the food contact legislation for rubbers is described in BfR (formerly BgVV) Recommendation XXI 'Commodity Articles Based on Natural and Synthetic Rubber'. There are separate requirements for silicone rubbers and these are contained within Recommendation XV.

3.4.1 Categories of Use

In the case of Recommendation XXI, four use categories and a special category are defined as follows:

- **Category 1** (Test conditions: 10 days at 40 °C). Rubber articles which come into contact with food for periods of more than 24 hours to several months, e.g., storage containers, container linings, seals for cans and bottles etc.

- **Category 2** (Test conditions: 24 hours at 40 °C). Rubber articles which come into contact with food for not more than 24 hours, e.g., food conveying belts, tubes and hoses, sealing rings for cooking pots, lock seals for milk can lids etc.

- **Category 3** (Test conditions: 10 minutes at 40 °C). Rubber articles which come into contact with food for not more than 10 minutes, e.g., milk liners and milking machine tubes, roller coatings and conveyor belts (fatty foods only in both cases), gloves and aprons for food handling etc.

- **Category 4** (No migration testing required). Rubber articles which are only used under conditions where no migration into food is to be expected, i.e., if the article comes into contact with the food for a very short time or only over a very small area and if it is not covered by Categories 1, 2 or 3. Examples of rubber products in this category include: conveyor belts, suction and pressure hosing for moving and loading/unloading dried food; tap washers, bevel seat valves, pump parts and other articles associated with the supply of drinking water.

- **Special Category** (Test conditions: 24 hours at 40 °C). Rubber articles directly associated with the consumption of food and which are being, or are expected to be, taken into the mouth, e.g., toys according to Recommendation XLVII, teats, soothers, gum shields, balloons etc.

The following food simulants are used in connection with the German regulations: distilled water, 10% ethanol and 3% acetic acid. The permissible migration limits vary according to the category and simulant.

The BfR regulations also include a number of specific composition and migration limits. For example:

1. *N*-Nitrosamines and nitrosatable substances

2. Amines (all categories)

3. Milking liners and milking tubes

4. Formaldehyde

5. Acrylonitrile

6. Zinc dibenzyldithiocarbamate

Pysklo and co-workers have compared the BfR list of ingredients approved for food contact rubbers with the equivalent Polish list. The Polish list was originally based on the German list and this exercise, carried out in 2002, showed that it was in need of updating [9].

3.4.2 Silicone Rubbers

Recommendation XV of the BfR regulations covers silicone oils, silicone resins and silicone rubbers. The section on silicone rubbers stipulates acceptable starting materials and the additives that may be used in processing and manufacture – both types and maximum levels. Separate restrictions are stated where silicone rubber is to be used for teats, dummies, nipple caps, teething rings or dental guards. Dummies and bottle teats must also comply with the requirements laid down in the Commodities Regulation (Bedarfsgegenstandeverordnung). The amount of volatile organic material is restricted to a maximum of 0.5%, as is the total extractable material. Test methods are referenced for these determinations as well as a test for residual peroxides which should be negative.

3.5 Other European Legislation

3.5.1 Requirements in France

French requirements for food contact elastomers (excluding silicones) are given in the Arrete of November 9th 1994 which is published in the Journal Officiel de la Republique Francaise, December 2nd 1994 pages 17029-17036 [10]. Four use categories (A to D) and a special category (designated T) are described together with a positive list detailing permitted ingredients in each category. There is an overall migration limit set at 10 mg/dm^2 (60 mg/kg), the same as for plastics. Other specific restrictions also apply, such as an SML for primary and secondary aromatic amines of <1 mg/kg.

3.5.2 Requirements in the Netherlands

These regulations, which closely resemble the CoE rubber resolution, can be found in Verpakkingen en gebruiksartikelenbesluit (Warenwet), Chapter III. There are positive lists of approved additives and food contact rubber products are divided into three parts [11].

3.5.3 Requirements in Italy

Italian requirements are given in the decree of March 21st 1973 contained with the supplemento ordinario alla *Gazzetta Ufficiale della Repubblica Italiana*, April 20th 1973 pages 12 to 14. There have since been updates, including the decree of June 3rd 1994.

3.5.4 Requirements in the United Kingdom

The UK legislation on food contact materials is published as a number of Statutory Instruments which were published in 1978 and came into operation in November 1979.

The use of rubbers in contact with food is covered by the legislation included in Statutory Instrument 1987 No. 1523 'Materials and Articles in Contact with Foodstuffs'. This states that any food contact material should not be injurious to the health of the consumer and that any contamination should not have an adverse effect on the organoleptic properties of the food (i.e., taint and odour).

The absence of any positive lists for compounding ingredients means that UK rubber compounders normally refer to either the FDA or the BfR regulations depending on the market to be addressed.

There are separate rules for the use of rubber in contact with potable water. These are given in the United Kingdom water fitting bylaws scheme and include tests for the following:

a) Taste [12, 13, 14]

b) Appearance [15]

c) Growth of aquatic microorganisms [16]

d) Migration of substances that may be of concern to public health [17]

e) Migration of toxic metals [18]

The test methods for the above are given in BS 6920, Parts 1 to 4 (2000-2001) [12-18].

4 Assessing the Safety of Rubber as a Food Contact Material

4.1 Special Considerations When Using Rubber as a Food Contact Material

The unique properties of rubber that make it such a useful material in food contact situations can also be the cause of some of its problems. The rubber matrix is lightly crosslinked and above room temperature the polymer chains are very mobile. So, unlike some of the other food contact materials (e.g., metals), the migration of low molecular weight compounds in and out of the material is relatively easy. Food can therefore penetrate into rubber and leach out the (potentially) large range of low molecular weight species within. It is possible to limit this ingress for particular applications by a careful choice of the rubber-food type combination. Although this matching process applies to the type of rubber, it is also important to take into consideration the complete rubber compound as up to 50% of a compound can be comprised of additives (e.g., plasticisers and fillers) and these can also have a profound effect on suitability and, hence, performance since if a rubber becomes swollen with the food stuff that it is in contact with a number of its important physical properties (e.g., hardness, resilience, tensile strength) will be affected. Although it is possible to make some generalisations (e.g., silicone rubber performs poorly in contact with fatty foods, and high levels of ester type plasticisers are also to be avoided with these), in common with all the other food contact materials, it is the contact conditions (time, area and temperature) that also play an important role in dictating ultimate suitability. In practice these interrelationships are well understood and form part of the body of knowledge used to draft food contact documents such as the CoE Rubber Resolution.

Another issue surrounds the stability of rubbers. The relatively reactive nature of rubber materials due to, either sites of unsaturation in the case of the diene rubbers (e.g., natural rubber, nitrile rubber and SBR), or large numbers of aliphatic hydrogen atoms, dictates that antidegradants are essential in most formulations if the polymer molecules and hence the materials' properties are to be protected. In addition to adding further to the list of potential migrants, this reactivity places restraints on the use conditions of different classes of rubber. **Table 4** provides a guide to the maximum service temperatures for a range of rubber types.

It is also the case that, in contrast to more modern materials such as thermoplastics, the technology associated with rubber has been developed over a long period. Most of this time predates the relatively recent period in which health and safety concerns have had such a profound influence on manufacturing practice and research and development

Table 4. Maximum service temperatures for a range of rubbers	
Rubber Type	**Maximum service temperature (°C)**
Natural rubber and polyisoprene rubber	80
Butadiene rubber and styrene-butadiene rubber	100
Nitrile rubber and polychloroprene rubber	120
Butyl rubber	130
Ethylene-propylene rubber	140
Chlorosulfonated polyethylene	150
Hydrogenated nitrile rubber	160
Acrylic rubber	160
Silicone rubber	225
Fluorocarbon rubber	250

activities. Hence, a number of the chemicals which have found widespread use for many years in rubber have become the subject of increased scrutiny to ensure they are suitable in today's climate.

4.2 Migration Tests

4.2.1 Overall Migration Tests

The aim of overall migration tests is to determine if a rubber is suitable for a particular food contact application. The methodology of the test varies depending on the regulations being addressed as does the way of expressing the data and the limits that have to be met. Some of the practical details of the methodologies associated with the different regulations are given in Section 3, with a further overview being given below.

4.2.1.1 FDA Regulations

Test pieces are cut from the rubber test product to provide a known surface area (cut edges are included in the calculation) and immersed in an appropriate amount (e.g., 100 ml) of food simulant (either hexane or distilled water). The samples are refluxed for seven hours in pre-cleaned glassware and then removed and placed into fresh simulant and refluxed for a further two hours. The test pieces are then removed and both the seven and two hour test portions evaporated separately to dryness in conditioned crucibles and the residues weighed. Blank determinations on equivalent volumes of the food simulant used are also performed. In order to be acceptable for food use the rubber has to pass the requirements given in Section 3.3.

4.2.1.2 BfR Regulations

Three food simulants are used in the BfR regulations (distilled water, 10% ethanol and 3% acetic acid) and the contact conditions for the four different food use categories for which migration testing is required are given in Section 3.4. Test pieces of 50 mm × 50 mm to give a total area of 50 cm² (both surfaces) are immersed in 100 ml of the appropriate simulant for the intended end use, the test performed and then the simulant dried down quantitatively.

The BfR limits are given in **Table 5**.

Table 5. Overal migration limits for the BfR regulations				
	Category (mg/dm²)			
	1	**2**	**3**	**Special**
Distilled water	50	20	10	10 or 50#
10% ethanol	50	20	10	-
3% acetic acid	150 (50)	100 (20)	50 (10)	-
(value) = permissible organics within total *#dependent on product type* *NB : no migration limit for Category 4 (see Section 3.4)*				

4.2.1.3 CoE Resolution

This has an overall migration limit of 60 mg/kg of food or food simulant for rubber products that are in Categories I and II (see Section 3.2). The choice of food simulant and the conditions that are used for the overall migration experiment (i.e., time and temperature) should be appropriate bearing in mind the conditions that the rubber product will see in service. Guidance for the designing of these tests is given in Technical Document No. 4 of the Resolution.

4.2.2 Specific Migration Tests

These tests are used to target specific chemical compounds for which there is a toxicological concern. The tests specified vary according to the regulations that are being studied, but some species (e.g., nitrosamines and nitrosatable substances) appear regularly due to the degree of concern associated with them. Other popular specific migrants include:

- Aromatic amines

- Other amines (e.g., cycloaliphatic amines)

- Peroxides and their breakdown products

- Formaldehyde

- Monomers (e.g., acrylonitrile)

- Accelerators (e.g., ZDBC, CBS, MBT)

These lists are not complete as it is recognised that rubber contains two important ingredients (antidegradants and curatives) that are reactive and so produce reaction and breakdown products. Recent work carried out at Rapra for the FSA (FSA contract number A03038) has shown that there are more than 1000 of these products originating from the 200 compounds on the CoE rubber resolution inventory list.

4.3 Fingerprinting Potential Migrants from Rubber Compounds

4.3.1 Use of Gas Chromatography-Mass Spectrometry (GC-MS) to Fingerprint Food Contact Rubber Samples

It is often useful to produce a qualitative or semi-quantitative fingerprint of the low molecular weight species in a rubber compound that have the potential to migrate into food. The technique of gas chromatography-mass spectrometry is often used for this due to its high resolution, important with rubbers due to their complexity, and the identification power of the mass spectrometer. In order to obtain data on as large a range of species as possible, it is often advisable to use both headspace GC-MS (the solid rubber samples being heated to a temperature of around 150 °C) and solution GC-MS where the solution is an extract of the rubber produced using a non-selective solvent (e.g., acetonitrile or acetone). Semi-quantitative data can be obtained by use of a single calibrant compound such as eicosane. Work using typical food contact rubber compounds has shown that, on average, between 20 and 30 compounds can be detected by conventional GC-MS using this approach [19]. In addition, the commercialisation of two dimensional GC-MS instruments has provided the analyst with greater resolving power, coupled with improved detection limits and enhanced deconvolution software, and this has increased this number to over 100 [5].

Two-dimensional GC-MS is of considerable interest to workers in the food contact field and a summary of the data that is contained within the Forrest paper is provided below.

For the purpose of comparing the data obtained using the two types of GC-MS technique, five high performance rubbers were compounded, all with food contact type applications in mind. These rubbers were therefore representative of the types of rubber that often require detailed profiling of the low molecular weight, potentially migratable, compounds. The formulations of the five rubbers are given in section 4.3.1.1:

4.3.1.1 Rubber Formulations

1. Nitrile Rubber - A peroxide cured hydrogenated nitrile rubber

Ingredient	phr
Zetpol 2000L	100
HAF N330	35
Zinc oxide	3
Perkadox 14/40	6
Antioxidant 2246	1

Notes

Zetpol 2000L = hydrogenated acrylonitrile-butadiene copolymer (nitrile) rubber

HAF N330 = carbon black filler

Perkadox 14/40 = 1,3-bis (*tert*-butyl-peroxy-isopropyl) benzene (curative)

Antioxidant 2246 = 2,2´-methylene-bis-(4-methyl-6-*tert* butylphenol) (antidegradant)

2. Fluorocarbon Rubber - A peroxide cured fluorocarbon tetrapolymer

Ingredient	phr
Viton GBL 200	100
Zinc oxide	3
MT N990	30
TAC	3
Luperco 101XL	4

Notes

Viton GBL 200 = Peroxide crosslinkable fluorocarbon rubber (composition undisclosed by the manufacturer, DuPont)

MT N990 = carbon black filler

TAC = triallyl cyanurate (curative)

Luperco 101XL = 2,5-dimethyl-2,5(di-*tert*-butyl-peroxy) hexane (curative)

3. Acrylic rubber

Ingredient	phr
Vamac G	100
Stearic acid	1.5
FEF N550	50
Diak No. 1	1.5
DOTG	4
Antioxidant 2246	1

Notes

Vamac G = ethylene-methyl acrylate (Acrylic) rubber

FEF N550 = carbon black filler

Diak No. 1 = hexamethylene diamine carbamate (curative)

DOTG = N,N´-di-ortho-tolyl guanidine (curative)

Antioxidant 2246 = 2,2´-methylene-bis(4-methyl-6-*tert*-butylphenol) (antidegradant)

4. Epichlorohydrin rubber

   General Description: An amine cured Hydrin rubber

Ingredient	phr
Hydrin 200	100
FEF N550	50
Maglite DE	7
Zinc oxide	3
Diak No. 1	1.5
Antioxidant MBI	2

   Notes

   Hydrin 200 = epichlorohydrin-ethylene oxide copolymer (Hydrin) rubber

   FEF N550 = carbon black filler

   Maglite DE = magnesium oxide

   Diak No. 1 = hexamethylene diamine carbamate (curative)

   Antioxidant MBI = 2-mercaptobenzimidazole (antidegradant)

5. Butyl rubber - A sulfur cured butyl rubber

Ingredient	phr
Butyl 268	100
Zinc oxide	3
Stearic acid	1
Vistanex LM-MS	10
ISAF N220	50
Sulfur	1
Rhenogran MPTD 70	1.43
DOTG	0.3
Nocrac AW	1

   Notes

   Butyl 268 = a isobutylene-isoprene (butyl) rubber

   Vistanex LM-MS = low molecular weight butyl rubber process aid

   ISAF N220 = carbon black filler

   Rhenogran MPTD 70 = DMDPTD = dimethyldiphenyl thiuram disulfide (curative)

   DOTG = *N,N*′-di-orthotolyl guanidine (curative)

   Nocrac AW = 6-ethoxy-1,2-dihydro-2,2,4-trimethylquinoline (antidegradant)

4.3.1.2 Experimental Conditions

The five compounds from Section 4.3.1.1 were analysed using the following methodology and experimental conditions.

Preparation of Solvent Extracts

A sample (0.3 g) was cut up finely and placed into a vial with 2 ml of acetone and subjected to ultrasonic agitation for 30 minutes. At the end of this period the extract was removed and placed into a GC sample vial. Analysis of the samples was carried out by both conventional GC-MS and two-dimensional GC-MS (GCxGC) using the following conditions:

GCxGC-Time-of-flight mass spectrometry (TOFMS) conditions

Instrument: Agilent 6890 Gas chromatograph with LECO Pegasus III GCxGC –TOFMS

Injection: PTV injection, 10 °C above primary oven, 1 μl

Primary column: J and W scientific HP-5MS 30 m × 0.250 mm, 0.25 μm film thickness

Secondary column: SGE BPX-50 1.8 m × 0.100 mm, 0.10 μm film thickness

Carrier gas: helium, 1.0 ml/min, constant flow

Primary oven program: 40 °C for 10 min, 10 °C/min to 320 °C, held for 15 min

Secondary oven program: 50 °C for 10 min, 10 °C/min to 330 °C, held for 15 min

Modulator offset: 30 °C

Modulator frequency: 4

Hot time: 0.30

MS 30-650 u, 76 spectra/s

Conventional GC-MS conditions:

Instrument: Agilent 6890/5973 GC-MS

Injection: 1 μl Splitless @ 310 °C

Carrier: Helium @ 1.0 ml/min

Column: Restek RTX-5 Amine 30 m × 0.32 mm, 1.0 µm film

Column oven: 40 °C for 5 min; 20 °C/min increase up to 300 °C, held for 15 min (30 min run time)

MS: Scanning 35 up to 650 Daltons every 0.25 s

To illustrate the differences obtained on the sample extracts by the GCxGC-TOFMS and GC-MS methods, the results obtained on one rubber, the hydrogenated nitrile rubber compound (Compound 1 above), are shown below in **Figures 1** and **2**.

It can be seen by a comparison of these two chromatograms that the two-dimensional GC-MS technique provides a much higher degree of resolution and hence separation of the species present in such complex samples. This was the case with all five compounds, the analytical power of the GCxGC-TOFMS technique resulting in a greater number of species being detected and identified in the acetone extracts. In addition to food contact work, this is very useful in reverse engineering projects where the degree of success is dependent on a good set of diagnostic compounds. These are defined as compounds that result from known breakdown products of additives such as curatives and antidegradants, as well as the original ingredients themselves, but does not encompass non-specific (e.g., general aliphatic and aromatic hydrocarbons) and other common compounds such as siloxanes.

A comparison of the approximate number of diagnostic compounds found in the acetone extract of each rubber compound by the two GC-MS techniques is shown in **Table 6**.

It can be seen from the data in the table that in most cases the two-dimensional GC-MS system has identified significantly more diagnostic species than conventional GC-MS.

In terms of comparing the two techniques for profiling all of the low molecular weight compounds present in rubbers, **Table 7** gives the total number of compounds detected and, of these, how many have been positively identified.

It can be seen from the data present in **Table 7** that the two-dimensional technique has produced a more extensive list of compounds and so offers greater advantages in the following type of work:

a) Profiling rubber compounds to assess the low molecular weight compounds that have the potential to migrate into food and pharmaceutical products.

b) Analysis of food and drug products which have contacted rubber components to determine if any species have migrated into them.

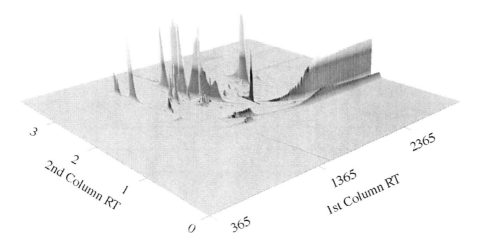

Figure 1. Three dimensional representation of the GCxGC-TOFMS TIC (total ion chromatogram) obtained for the acetone extract of the hydrogenated nitrile rubber

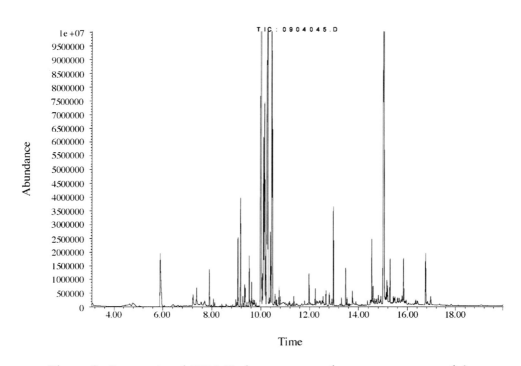

Figure 2. Conventional GC-MS chromatogram for acetone extract of the hydrogenated nitrile rubber sample

Table 6. Comparison of the number of diagnostic species		
Compound	GCxGC-ToFMS	Conventional GC-MS
Nitrile	>35	>30
Fluorocarbon	>20	>10
Acrylic	>30	>20
Hydrin	>10	>10
Butyl	>15	>5

Table 7. Comparison of the total number of species detected		
Compound	GCxGC-ToFMS	Conventional GC-MS
Nitrile	102 (53)	66 (32)
Fluorocarbon	184 (118)*	36 (22)
Acrylic	217 (79)	55 (29)
Hydrin	46 (23)	37 (20)
Butyl	111 (41)	27 (15)

( ) = *Compounds for which either a match on the mass spectral database has not been found or the match is tentative*

* *the number of unassigned peaks for this compound was increased by virtue of the fact that some of the main breakdown products of the peroxide used (e.g., dimethyl hexanedione) were not in the mass spectral data base. These aliphatic ketones also give very simple spectra (i.e., with only two or three diagnostic ions) and so positive manual identification is difficult; standards would need to be run to confirm their presence using detention time*

Table 7 also shows that the proportion of compounds for which no assignment, or only a tentative assignment, can be found is roughly the same for the two approaches.

4.3.2 Use of Liquid Chromatography-Mass Spectrometry (LC-MS) to Fingerprint Food Contact Rubber Samples

As in-house databases are developed as a result of using liquid chromatography-mass spectrometry (LC-MS) on rubber samples, the inclusion of this technique into the fingerprinting process will complement GC-MS data by contributing information on thermally labile and relatively large (e.g., oligomeric) potential migrants. Acetone, acetonitrile and diethyl ether solutions of rubber extracts can all be analysed directly by LC-MS. This has been demonstrated recently in a paper by Sidwell, [20] which describes how LC-MS was used to provide additional information on the species present in an ether extract of a food contact EPDM rubber. The formulation of the rubber compound used (designated EPDM2) is given in Section 4.5.2.

Figure 3 shows the total ion current GC-MS trace for the examination of the diethyl ether extractable species from the sulfur cured EPDM2 compound. The species detected in this chromatogram are shown in **Table 8**. It is important to note that these species are all of relatively low molecular weight.

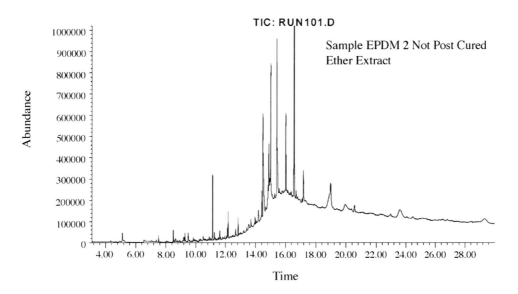

Figure 3. GC-MS chromatogram of the diethyl ether extract of the EPDM2 sample

In an attempt to obtain more information on this diethyl ether extract, it was also analysed by LC-MS using a C_{18} reverse phase gradient elution separation with atmospheric pressure chemical ionisation (APCI). Using the APCI head in the positive mode gave the total ion current trace shown in **Figure 4**. The last three peaks showed ions of masses 538, 566 and 594 and are believed to relate to the presence of tellurium dithiocarbamates in the extract (ions $+2H^+$ from the ionised protonic solvent).

Table 8. Species extracted from EPDM2 and detected by GC-MS (Figure 3)		
Peak time (mins)	Assignment	Origin
5.11	Ethane, isothiocyanato-	From cure system
7.52	Dicyclopentadiene	Monomer
8.50	*N*-formylpiperidine	From cure system
9.19	Tetramethylthiourea	From cure system
9.27	Benzothiazole	From cure system
11.92	Benzothiazole, 2-(methylthio)-	From cure system
12.16	2-Benzothiazolamine, *N*-ethyl-	From cure system
14.48	Dodecanamide, *N*-(2-hydroxyethyl)-	From process aid
14.88	Pyrene	From carbon black
15.09	*N,N*-dimethylpalmitamide	From process aid
15.42	Dodecanamide, *N*-(2-hydroxyethyl)-	From process aid
16.57	Di(2-ethylhexyl)phthalate	Contaminant
17.18	Thiazole, 4-ethyl-2-propyl-	From cure system

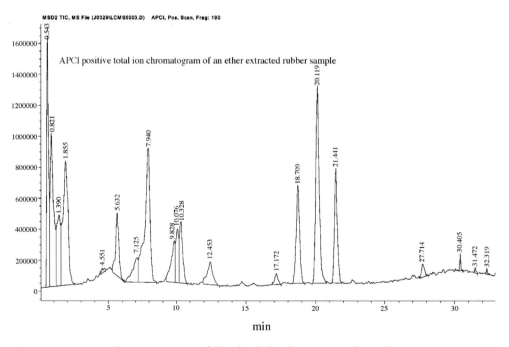

Figure 4. LC-MS chromatogram of the diethyl ether extract from the EPDM2 sample

- 538 = tellurium^{4+} with two dimethyl dithiocarbamate groups and two diethyldithiocarbamate groups (tellurium^{4+} being detected separately)

- 566 = tellurium^{4+} with three dimethyl dithiocarbamate groups and one diethyldithiocarbamate group

- 594 = tellurium^{4+} with four diethyldithiocarbamate groups

Clearly the presence of such high molecular weight species in this extract was not detected by the initial GC-MS work.

4.4 Determination of Specific Species in Rubbers and Migrants in Food Simulants and Food Products

The national regulations and the CoE resolution stipulate concentration limits for certain species within food contact rubber compounds (Section 3) and they also have SML for certain migrant compounds.

Analysis work is therefore required on a quality assurance basis to ensure that a food approved rubber compound remains fit for purpose (for example by checking the monomer level) and to ensure that compounds having SML do not exceed them in food simulant or food samples prepared using appropriate contact conditions.

The main classes of potential migrants, and the analytical techniques that are used to detect and quantify them, are described next.

4.4.1 Monomers

Monomers are either gaseous or relatively volatile liquids and so GC and GC-MS based techniques are used to determine them in both the rubber compounds and food simulants/food products. To simplify the analysis, a static headspace sampler is often used to isolate the monomer from the sample matrix; an extraction procedure presenting chromatographic problems with the extraction solvent and co-extractants often obscuring the analyte.

4.4.2 Plasticisers and Process Oils

These additives are essentially viscous, high boiling point liquids and so the most appropriate technique to use is LC-MS. A range of synthetic plasticisers such as phthalates, adipates, mellitates and sebacates can be detected using APCI ionisation.

Process oils are hydrocarbon mineral oils and require either the APCI head (which can ionise non-polar species) or, where the oil contains sufficient aromatic character, the use of in-line UV or fluorescence detectors. A fluorescence detector is particularly sensitive in the detection of polycyclic aromatic hydrocarbon (PAH) compounds in such oils.

4.4.3 Cure System Species, Accelerators and Their Reaction Products

This class of additive can present problems as they are often thermally labile, reactive and, in some cases, have a degree of ionic character (e.g., dithiocarbamate salts). In some cases the reaction products (e.g., methyl aniline from DOTG and cyclohexylamine from CBS) are stable and so GC and GC-MS can be used. Peroxides are popular curatives for food use rubbers and the stable, breakdown products of these can be easily detected by GC-MS.

In many cases LC-MS is a more appropriate technique than GC-MS. It is also easier to use LC-MS with a number of the approved food simulants as they can be injected directly into the instrument, being compatible with the mobile phase.

Nitrosamines are derived from the secondary amines that are the breakdown products of a number of commonly used accelerators (see Section 5.1). These potentially carcinogenic species can be determined at low ppm levels by the use of a combined gas chromatography-thermal energy analyser instrument. Samples can be prepared from rubber compounds by either extraction or food migration studies and then, after a concentration step, injected into the gas chromatograph. The separated nitrosamines enter a catalytic pyrolyser where nitrosyl radicals are generated. These react with ozone introduced into the system to form a new radical which is chemiluminescent as it returns to the ground state. The emitted light generated by this loss of energy is detected and quantified. A number of factors have been found to affect the levels of nitrosamines found in a particular rubber sample (see Section 5.1).

4.4.4 Antidegradants and Their Reaction Products

This class of additive is less thermally labile and reactive than the preceding one and gas chromatography based methods can be used for a number of the members. However, due to the relatively large number of high temperature processing steps used with rubbers (i.e., mixing, extruding/calendaring, moulding and, in some cases, postcuring) a number of low volatility, oligomeric antidegradants are commercially available and the higher oligomers of these far exceed the molecular weight limits of GC. LC-MS methods therefore have to be used and the technique has proved to be of great value in determining a range of amine and phenolic types.

4.4.5 Oligomers

The wide availability of LC-MS instruments means that they are now rivalling supercritical fluid chromatography (SFC) for the analysis of oligomers. The range of LC-MS instruments can be extended up to 4000 daltons and this capability makes them ideal to characterise oligomers. For example, it has been shown that silicone oligomers can be detected by this route in food simulants (FSA contract A03046).

4.5 Research Studies Carried out at Rapra for the FSA

A number of research projects have been commissioned by the UK Government's FSA to study the effects that rubber has on food. Four projects of this kind have been carried out by Rapra Technology over the past 10 years. An overview of each of the four projects is given below in Sections 4.5.1 to 4.5.4.

4.5.1 FSA Project FS2219 – Migration Data on Food Contact Rubbers

4.5.1.1 Introduction

This project ran from 1994 until 1997 and was one of the first major projects in the UK to look at the use of rubber in the food industry. During this three year period, the project achieved the following:

1. Reviewed the use of elastomeric materials throughout the UK food industry

2. Developed suitable methods of analysis and undertook an in-depth examination of the types and likely levels of migratory components from five specially prepared rubber compounds representative of those that are used in contact with food

3. Examined migration from a range of rubber materials used in contact with foodstuffs as supplied by an industrial group who also participated in the project

4. Appraised the scientific basis of national approaches to food contact legislation in the work programme in order to move forward with the ultimate development of unified regulations

5. Examined migration through development of a range of analytical procedures and protocols which have addressed the important issues relevant to applications

6. Provided migration data obtained to the Committee of Experts of the Council of Europe to assist with their discussions on the writing of resolutions concerning food contact elastomers.

4.5.1.2 Standard Rubber Compounds

One of the important objectives of this project was to establish the types of rubber compounds that were used in all areas of the food industry and to correlate data on contact times, contact temperatures and contact areas. This data, together with views and advice submitted by industry, was used to develop a number of standard rubber compounds which were then used to develop a range of analytical procedures and protocols, leading to the generation of migration data (both overall and specific) using food simulants.

The following guidelines were used in the designing of the rubber compounds:

a) The formulations should be representative of those used in contact with food

b) The specific compounding ingredients were either on the BfR or the USA FDA food contact inventory lists

c) The rubber types chosen reflected the range most commonly used with food

d) A range of accelerated sulfur and peroxide cure systems were chosen to mimic the diversity present in food contact compounds

e) Two compounds contained cure systems that were known to generate nitrosamines

f) One compound included a process oil which would provide a source of polycyclic aromatic hydrocarbons

g) The same grade of carbon black was utilised in the four black compounds for consistency

h) Hydrocarbon waxes and a hydrocarbon (paraffinic) oil were included in certain compounds so that potential hydrocarbon migration could be studied.

The formulations of the five compounds that were produced are shown next:

Natural rubber compound (NR 1) - typical formulation for hosing and belting (all phr)

Natural rubber (SMR L grade)	100
Zinc oxide	5
Stearic acid	1
Naphthenic oil	5
4,4′-thio-bis(3-methyl-6-*tert*-butylphenol	1
6PPD	1

Hydrocarbon wax	2
SRF N762 Carbon black	10
Calcium carbonate	60
Sulfur	1
CBS	2.5

Nitrile rubber compound (NBR 1) – typical flexible food contact compound (all phr)

Nitrile rubber (34% acrylonitrile)	100
Zinc oxide	5
Stearic acid	1
Di-2-ethylhexyl phthalate	10
4,4′-thio-bis(3-methyl-6-*tert*-butylphenol	1
IPPD	1
Hydrocarbon wax	2
SRF N762 Carbon black	30
Clay	20
Sulfur	1.5
TMTD	0.25
MBTS	1

Ethylene-propylene compound (EPDM 1) – high temperature sealing gasket formulation (all phr)

EPDM	100
Zinc oxide	5
Paraffinic oil	8
SRF N762 Carbon black	50
1,3-bis(*tert* butylperoxyisopropyl) benzene	4

Fluorocarbon rubber (FKM 1) – typical high temperature (e.g., 200 °C) use compound (all phr)

Vinylidene fluoride hexafluoropropylene copolymer	100
Magnesium oxide	15
SRF N762 Carbon black	20
Carnauba wax	0.5
Hexamethylene-diamine carbamate	1.25

Silicone rubber (Sil 1) – general purpose compound

Polydimethyl vinyl siloxane (vinyl content ~0.5%) containing bis(2,4-dichlorobenzoyl) peroxide (50% active) at a level of 1.2%.

This was a pre-compounded material which contained fumed silica and process aids to give a Shore A hardness of 60.

A wide variety of tests were conducted on the above compounds. Some of these tests were designed to determine the level of migratable species present in the compounds, others were to determine the specific identity of these species. Accurate quantification work was also carried out on specific groups of species, e.g., nitrosamines.

4.5.1.3 Migration Experiments Carried Out on the Standard Rubber Compounds

Overall Migration Tests

Overall migration tests were performed on the compounds as prescribed under both the FDA and BfR regulations. In the case of the FDA test, both hexane and distilled water food simulants were used (see Section 3.3), and work to the BfR guidelines involved all three simulants using the contact conditions applicable to Categories 1, 2 and 3 (see Section 3.4).

Specific Migration Tests

The following specific determinations have been made on the BfR test simulants (excluding the examination of the EPDM and silicone rubber extracts for amines and N-nitrosamines as these species were not present in these compounds):

1. Nitrosamines

2. Primary aryl and/or secondary N-alkyl-arylamines

3. Secondary aliphatic and cycloaliphatic amines

4. Formaldehyde

5. Dichlorobenzoic acid – silicone compound only (breakdown product of the peroxide curative).

Migration Tests Using GC and GC-MS

In addition to the overall and specific migration tests, a number of analytical methods were used that employed GC and GC-MS. These were:

1. Overall migration tests on food simulants (10% ethanol, 95% ethanol and isooctane) using quantification by GC.

2. Determination (i.e., fingerprinting) of potential migrants by dynamic headspace GC-MS

3. Determination (i.e., fingerprinting) of potential migrants by analysing diethyl ether extracts by GC-MS

4.5.1.4 Results of the Migration Experiments

Summary of the Key Findings for the FDA and BfR Extraction Tests on the Five Rubber Compounds

1. Natural rubber compound

   The natural rubber compound did not meet the extraction requirements of the FDA for use with fatty foods, but did meet the requirements for aqueous foods. It complied with all the BfR requirements for Category 3 materials, but not for Categories 1 and 2, where the limits were exceeded for acidic foods, and also *N*-nitrosomorpholine extraction into all three simulants in the 24 hour and 10 day tests.

2. Nitrile rubber compound

   The nitrile rubber compound did not meet the extraction requirements of the FDA for use with fatty food. All the BfR extraction requirements were met for Category 1 to 3 materials.

3. EPDM rubber compound

   The EPDM compound complied with FDA extraction requirements for aqueous and fatty food. The BfR extraction requirements were met for Category 2 and 3 materials, but it failed the requirements for Category 1 materials contacting acidic foods for greater than 24 hours due to the amount of formaldehyde detected in 3% acetic acid after 10 days at 40 °C.

4. Fluorocarbon rubber compound

   The fluorocarbon rubber complied with the FDA extraction requirements for use with aqueous and fatty foodstuffs. The compound complied with the BfR Category 3 material requirements, but failed the Category 1 and 2 requirements due to the high level of extractables in 3% acetic acid. These extractables were mainly magnesium acetate formed due to a reaction of the acid with the magnesium oxide.

5. Silicone rubber compound

   The silicone rubber complied with the FDA extraction requirements for aqueous and fatty foods. It failed the BfR requirements for Category 1 (10 day test) for extractable formaldehyde with the 3% acetic acid and the 15% ethanol simulants.

Summary of the Specific Constituents Migrating from the Five Rubber Compounds in the BfR Tests

When the rubber compounds were examined using the specific testing procedures in the BfR tests, the following compounds were detected:

a) Natural rubber compound

Amines	Isopentylamine
	Dimethylamine
	Isopropylamine
	Diethylamine
	Morpholine
N-nitrosamines	N-nitrosodimethylamine
	N-nitrosodiethylamine
	N-nitrosomorpholine
	N-nitrosodibutylamine
Aldehyde	Formaldehyde

b) Nitrile rubber compound

Amines	Isopentylamine
	Dimethylamine
	Isopropylamine
	Diethylamine
	Methylamine

N-nitrosamines	N-nitrosodimethylamine
	N-nitrosodiethylamine
	N-nitrosodibutylamine

Aldehyde	Formaldehyde

c) EPDM rubber compound

Aldehyde	Formaldehyde

d) Fluorocarbon rubber compound

Amine	Hexamethylene diamine

e) Silicone rubber compound

Aldehyde	Formaldehyde

General Conclusions from the Migration Studies Carried out on the Rubber Compounds

The main conclusions that could be drawn from the migration studies carried out on the five compounds during the course of this project were:

1. Overall migration data obtained using appropriate simulants and test conditions were in accordance with national regulatory requirements, i.e., 10 mg/dm^2 or 60 mg/kg.

2. If tested with simulants under unsuitable conditions for the type of rubber compound, then high levels of overall migration into simulants may be observed.

3. The main types of specific migratory compounds, particularly in aqueous food simulants, are normally decomposition or reaction products of the vulcanisation system.

4. Levels of specific controlled migrants have generally been in compliance with national requirements in tests of up to and including 24 hour contact duration. Longer term testing of the compounds, particularly those cured with sulfur/sulfur containing accelerators (10 days at 40 °C) has in a number of cases given levels of extracted N-nitrosamines and extractable aromatic amines in excess of BfR limits.

5. Compliance with FDA extraction requirements for articles for repeated use in contact with fatty foods is difficult to achieve with compounds containing plasticisers or oil extenders.

A detailed and comprehensive listing of the overall and specific migration data values obtained on these five rubber compounds is given in the first Rapra Rubbers in Contact with Food Review Report [1].

4.5.2 FSA Project FS2248 – Further Migration Data on Food Contact Rubbers

4.5.2.1 Introduction

The objective of this project, was to build on the body of information provided by FS2219 by obtaining additional information on migrants from rubbers used in contact with food, and to the examine the effect of ageing, sanitisers and cleaning agents on the nature and levels of migrants. As with FS2219, an industrial group of companies participated in this project.

4.5.2.2 Standard Rubber Compounds

The rubber samples that were examined during the course of this project were selected in consultation with the FSA and industry. They included five new compounds and two (the nitrile rubber compound, NBR 1, and the EPDM 1 compound), which were used in the FS2219 project.

The five new compounds were formulated as follows:

Natural rubber compound (NR2)

Natural rubber (SMR CV 60)	100
Zinc oxide	5
Stearic acid	1
4,4′-Thio-bis(3-methyl-6-*tert*-butyl phenol)	1
N-(1,3-Dimethyl butyl)-*N*′-phenyl-*p*-phenylenediamine	1
Hydrocarbon wax (Okerin 1944)	2
HAF N330 Carbon black	35
Naphthenic oil	5

Sulfur	1
CBS	2.5

Notes: This compound is similar to NR1, with the calcium carbonate and SRF black replaced with HAF black.

Nitrile rubber compound (NBR2)

Nitrile rubber (Krynac 34/50)	100
Zinc oxide	5
Stearic acid	1
4,4′-Thio-bis(3-methyl-6-*tert*-butyl phenol)	1
IPPD	1
Hydrocarbon wax (Okerin 1944)	2
FEF N550 Carbon black	85
Di-2-ethyl hexyl phthalate	10
Sulfur	1.5
TMTM	0.5

Notes: A relatively hard seal compound containing a 'clean' black at a level above that which would be permitted by the FDA, but included to examine any potential migrants from the black, amongst other things.

EPDM rubber compound (EPDM 2)

EPDM (Keltan 720)	100
Zinc oxide	5
Stearic acid	1
SRF N762 Carbon black	50
Paraffinic oil (Strukpar 2280)	8
Struktol WB16	1
Sulfur	2
MBT	1.5
TMTD	0.8
TDEC	0.8
Dipentamethylene thiuram hexasulfide	0.8

Notes: This is a 'triple 0.8' cured EPDM, a type of system widely used for non-food use EPDM rubbers and the objective of including it was to see if this type of compound would be suitable for food use.

Fluorocarbon rubber compound (FKM 2)

Vinylidene fluoride-hexafluoropropylene copolymer	100
Calcium hydroxide	6
Magnesium oxide	3
Calcium metasilicate filler (Nyad 400)	40

Notes: The polymer (Viton E60C) was pre-compounded with the following curatives:

a) 1.65% of a masterbatch comprised of 33% benzl-triphenyl-phosphonium chloride/67% Viton polymer, and

b) *4% of a masterbatch of 50% 2,2-bis-4-hydroxyphenyl perfluoropropane/50% Viton polymer*

This compound is typical of a high quality gasket formulation. It was postcured for 24 hours at 200 °C in an oven after the initial press cure to conform to standard manufacturing practices.

Fluorocarbon rubber compound (FKM 3)

Viton A-401C Polymer	100
(pre-compounded with a bisphenol A based cure system)	
FEF N550 Carbon black	20
Calcium hydroxide	6
Magnesium oxide	3

Notes: DuPont state that pre-compounded Viton A-401C is in compliance with FDA regulation 21 CFR 177.2600. This compound was also postcured in an oven prior to testing.

4.5.2.3 Tests Carried Out on the Seven Rubber Compounds

The above five compounds, and the two (NBR 1 and EPDM 1) from the previous study, were examined for migratable or extractable species according to EC Framework Directive 89/109/EEC for food contact materials and national legislative schemes (see Section 3). Migration was studied before and after the samples had been treated with chemical sanitisers and cleaning agents, and also following ageing.

Tests carried out on the non-treated samples included:

a) Repeated distilled water contact - to obtain information on potential migrants into aqueous food

b) Diethyl ether extraction – for information on potential migrants, particularly into fatty food

c) Examination of volatile species – for information on potential high temperature migrants

Other tests that were undertaken were designed to examine the nature and amount of potential migrating or extractable species following:

a) Autoclaving

b) Sodium hypochlorite (pH 7 and alkaline) sanitation treatment

c) Sanitation with didecyl dimethyl ammonium chloride

d) Gamma radiation sterilisation

e) Cleaning with a nitric/phosphoric acid descaler

f) Cleaning with an alkaline cleaner

g) Heat ageing (particularly to see whether any new migrating species are formed)

Of particular interest was the amount and composition of water-soluble migrants from the rubber compounds following the above treatments, including any specific migrants that are likely to have a 'not detectable' or very low SML restriction in any future harmonised European legislation, e.g., aromatic amine, *N*-nitrosamines, polycyclic aromatic hydrocarbons etc.

Migration testing was mainly undertaken by contacting 1 dm^2 of rubber surface with 100 ml of simulant or sanitation/cleaning solution. Some analytical method development studies and 'round robin' testing of a specific migration screening procedure were carried out.

4.5.2.4 Results Obtained During the Course of the Project

A summary of the results obtained, and conclusions drawn, during the course of this project is shown below:

1. Potential migrants detected in the compounds – general comments

a) All of the rubber compounds were found to contain chemical species that have the potential to migrate into food, but whether they will be of concern will depend upon the time, temperature and surface area of the rubber that is in contact with the food and its fat content.

b) The use of the CBS accelerator in compound NR2 led to relatively high levels of water soluble migrating species derived from this compound, such as mercaptobenzothiazole, benzothiazole and a number of other related substances (see **Figures 5** and **6**). As expected, the levels of these diminished upon repeat exposure of the rubber and after high temperature ageing. Similar chemical species would result from the use of other sulphenamide accelerators and this probably accounts for their absence from the BfR inventory list – despite being included in the FDA list.

It was also possible to use GC-MS to quantify the migration of the two major CBS breakdown products into water from compounds NR2 and EPDM2. The results obtained for mercaptobenzothiazole and benzothiazole are shown in **Table 9**.

Table 9. Migration of mercaptobenzothiazole and benzothiazole (mg/kg) into distilled water at 40 °C from NR2			
Migrant	1st 24 h extraction	2nd 24 h extraction	3rd 24 h extraction (carried out after autoclaving the 2nd 24 h extraction sample)
Sample NR2			
Mercaptobenzothiazole	17.6	6.5	13.9
Benzothiazole	1.2	1.0	Not detected
Sample EPDM2			
Mercaptobenzothiazole	1.0	1.4	2.8
Benzothiazole	0.5	0.3	Not detected

c) The 3 × 0.8 mixed accelerator system used in the EPDM2 compound produced a wide range of complex water soluble reaction products which would probably be a hindrance in obtaining its approval for food use.

d) Subjecting a diethyl ether extraction screening test for permitted additives to a 'round robin' assessment was successful in that similar GC-MS data was obtained by the participating laboratories, but revealed the complexities in interpreting such data as not all were able to relate the breakdown/reaction products detected to the original ingredients.

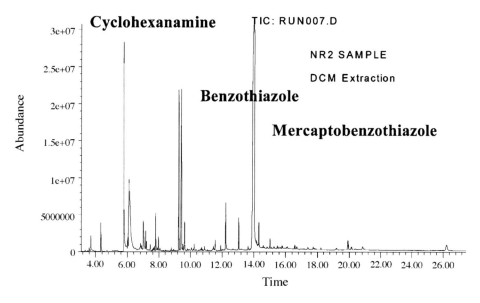

Figure 5. Water extractables (24 hours 40 °C) from Compound NR2 (partitioned into dichloromethane (DCM) and then examined by GC-MS)

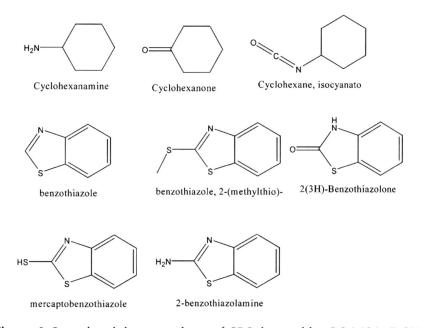

Figure 6. Some breakdown products of CBS detected by GC-MS in DCM partitioned water extracts (see Figure 5)

2. Potential migrants detected – specific species

a) Diethyl ether extraction data indicated that fatty foods could potentially extract organic additives such as plasticisers, extenders and antidegradants (unless of high molecular weight or chemically bound to the polymer chain) and also cure system reaction products.

b) The polycyclic aromatic hydrocarbon pyrene was present in the diethyl ether extract from two rubbers that contain SRF N672 carbon black (EPDM2 and NBR2). Pyrene and fluoranthene were present in the volatiles from EPDM2 on heating at 150 °C. Another PAH, 2-methylfluorene was detected in the water extract from NR2, which contained HAF N330 carbon black. These findings confirmed the requirement for using high purity carbon blacks in food contact rubbers.

c) The toxic monomer, acrylonitrile, was released from the NBR2 compound on heating at 150 °C. Dicyclopentadiene monomer was also released from the EPDM2 sample under similar conditions. It was unclear whether or not these monomers were present as polymerisation residues or were generated by the heating process. Migration testing for these monomers is advisable.

d) Migration of detectable quantities of aromatic amines from compounds containing *para*-phenylene diamine antiozonants has been observed.

e) The cure system breakdown product, triphenyl phosphine oxide, was a major migrant for all of the work undertaken with the two fluorocarbon rubbers (FKM2 and FKM3). The data on this migrant for the FKM3 compound is shown in **Table 10**. However, no bisphenol A was found from either of these rubbers.

Table 10. Levels of extractable triphenyl phosphine oxide from FKM3	
24 h at 40 °C in distilled water tests	Concentration
Irradiated	5.6 mg/kg as hydrocarbon
Aged 3 months at 40 °C	3.9 mg/kg as hydrocarbon
Aged 20 days at 150 °C	1.7 mg/kg as hydrocarbon
Other tests	
Acid cleaner 2 h at 85 °C (1st exposure)	23.4 mg/kg as hydrocarbon
Alkaline cleaner 2 h at 85 °C (2nd exposure)	13.3 mg/kg as hydrocarbon

3. Effect of ageing and sanitisers/cleaning agents on the potential migrants identified

a) Autoclaving samples does not necessarily reduce the amount of available migrating species (see **Table 9** where the level of mercaptobenzothiazole from NR2 and EPDM2 increases after autoclaving) and can lead to the production of new migrants.

b) Heat ageing at high temperatures, when it results in surface hardening, has reduced the level of migrating species. However new potential migrants can be formed (see **Figure 7**).

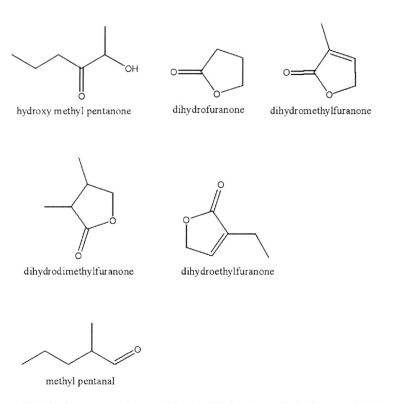

hydroxy methyl pentanone   dihydrofuranone   dihydromethylfuranone

dihydrodimethylfuranone   dihydroethylfuranone

methyl pentanal

Figure 7. Ethylene propylene rubber (EPDM2) aged 20 days at 150 °C - new species formed

c) Changes in the composition of migrants with sample storage was observed. Some of this was associated with the expected blooming to the surface of compounding ingredients such as antidegradants, other changes were due to further reactions of breakdown products resulting from the cure system additives.

d) Chemical sanitation with quaternary ammonium based sanitiser at 85 °C also resulted in the formation and extraction of chlorinated curative residues. The test temperature used resulted in relatively high levels of soluble migrants.

e) No new species specifically associated with gamma irradiation were observed.

f) The work with cleaning agents highlighted one potential problem with compounds that contain thiurams or dithiocarbamate accelerators. The acidic cleaner tested was found to generate nitrosamines due to reactions with it and the breakdown products of these accelerators (see **Figure 8**). However, levels of these compounds would drop with repeated contact in service.

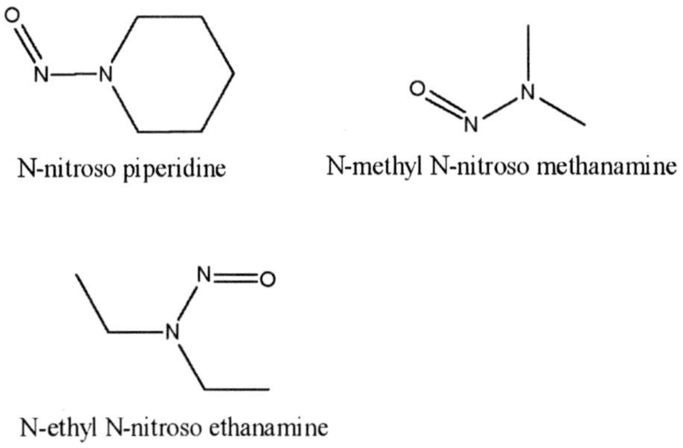

N-nitroso piperidine   N-methyl N-nitroso methanamine

N-ethyl N-nitroso ethanamine

Figure 8. Nitrosamines detected in the acid cleaner extract from EPDM2

Table 11. Major migrants in 6 day water soak (~23 °C) after alkaline hypochlorite treatment of NR2	
Species	Approximate concentration (mg/kg) expressed as hydrocarbon
Mercaptobenzothiazole	131
N-cyclohexyl formamide	29.8
Chlorinated benzothiazole	24.0
Benzothiazole	10.1
Benzothiazolone	7.2

g) Chemical sanitation with sodium hypochlorite can lead to the formation of chlorinated organic species from the rubber compound. For example, chlorinated benzothiazole, and various chloroanilines have been detected in the NR and NBR compounds studied (from the phenylene diamine antiozonants present) – see **Table 11**.

Extensive tabulated data on this project is present in the full project report which is available from the FSA library.

4.5.3 Project A03038 – Rubber Breakdown Products

4.5.3.1 Introduction

This project ran from 2002 until 2004 and was entitled 'An investigation of the breakdown products of curatives and antidegradants used to produce food contact elastomers'. Essentially the 161 curatives and antidegradants present in the Inventory List present in the CoE Resolution on Food Contact Rubbers were taken and all of the breakdown and reaction products of these listed during the course of the project. Products that resulted from the use of these additives individually as well as products that could result from interactions between the compounds were considered. Because of the chemical reactivity of these types of rubber additive and the low molecular weight of the breakdown products, this work was vitally important if all of the potential migrants in a particular rubber compound are to be accurately catalogued.

The salient aspects of the project were:

1. A literature search was conducted to establish the extent of the information that was already available on breakdown products

2. Prime examples of the blends of curatives that are used in industry to formulate food contact rubbers were listed

3. For those 161 CoE curatives and antidegradants where information was not available, theoretical predictions of the breakdown products that would result were made. Products resulting from the typical blends established in 2 were included.

4. To test the predictions made in 3, nineteen industrially relevant rubber compounds, covering all the main rubber types used in the food industry, were produced and analysed by GC-MS and LC-MS to fingerprint the species present. This was carried out before and after ageing (using representative conditions for each rubber) had been carried out on test sheets produced from the compounds.

A list of the compounds produced, showing the generic rubber types, is given next (the full formulation of each compound is shown in Appendix 1):

a) 20D   Amine cured Hydrin rubber

b) 48V   Amine cured fluorocarbon rubber

c) 49V   Peroxide cured fluorocarbon rubber

d) 50B   Sulfur cured butyl rubber – compound 1

e) 51B   Sulfur cured butyl rubber – compound 2

f) 195S   Sulfur cured SBR rubber

g) 201E   Polychloroprene rubber

h) 238C   Peroxide cured EPDM rubber – compound 1

i) 239C   Peroxide cured EPDM rubber – compound 2

j) 240C   Sulfur cured EPDM rubber – compound 1 ('triple 8' accelerator system)

k) 241C   Sulfur cured EPDM rubber – compound 2

l) 242C   Sulfur donor cured EPDM rubber

m) 349N   Peroxide cured hydrogenated nitrile rubber

n) 350N   Sulfur cured nitrile rubber – compound 1

o) 351N   Sulfur cured nitrile rubber – compound 2

p) 352N   Sulfur cured nitrile/SBR blend rubber

q) 503A   Sulfur cured natural rubber – compound 1
              (efficient vulcanisation system)

r) 504A   Sulfur cured natural rubber – compound 2

s) Vamac 8 amine cured acrylic rubber

The extent to which the breakdown products could migrate was established by carrying out migration work using food simulants and some food products. Analysis was carried out by both GC-MS and LC-MS.

The results obtained during the course of the project are summarised in Section 4.5.3.2.

4.5.3.2 Listing of the Breakdown Products for the CoE Curatives and Antidegradants

From the 161 curatives and antidegradants on the CoE Inventory list it was possible to predict that there would be at least 900 breakdown/reaction products. This total included products that would result from typical blends of curatives (e.g., thiurams with sulfenamides) used in sulfur based cure systems. Obviously, the number of blends included in such an exercise had a large influence on the total number of products listed.

All types of cure system (i.e., sulfur, peroxide, amine etc.) and antidegradants (i.e., phenolic, amine, phosphites, oligomeric etc.) were included in this exercise. The curatives and antidegradants present in the CoE Rubber Resolution list can be broken down as follows:

Curatives

Thiurams	8
Thioureas	3
Thiazoles	8
Sulfenamides	2
Xanthates	16
Peroxides	16
Guanidines	6
Dithiocarbamates	21
Amines	11
Others	9
Total	*100*

Antidedegradants

Phenolics	30
Amines	21
Phosphites	5
Thioesters	3
Others	2
Total	*61*

The situation was complicated by the fact that for a number of the classes of additives in the list certain members could have two completely different functions. For example:

- Xanthates and dithiocarbamates - Fast accelerators in sulfur cures and short stops in emulsion polymerisation recipes

- Peroxides - Curatives and polymerisation initiators

- Amines - Curatives, accelerators and antidegradants

Care was therefore taken to include those members of these groups that were known to be used industrially as either curatives or antidegradants, or both.

With respect to curative blends, by the use of a number of sources (e.g., published literature, text books, industrial contacts etc.,) it was possible to identify at least forty commercially important examples. The blends, all of them associated with sulfur cure systems, were used in a wide range of rubber types (e.g., NR, NBR, EPDM and butyl).

Table 12. Possible breakdown products for the nitrile compound 351N		
Ingredient	**Description**	**Breakdown/reaction products**
Breon ($N_{36}C_{60}$)	Nitrile Rubber	dodecenes
Stearic acid	Activator	amides of amines listed below
(mixture of C_{14}, C_{16} and C_{18} acids)		
DPG	Accelerator	N,N'-diphenylthiourea trisphenyl amino-1,3,5-triazine diphenylcarbodiimide aniline ammonia N-phenyl-N-methyl-N'-phenylthiourea N,N'-diphenylurea diphenylamine phenyl isothiocyanate phenyl isocyanate
MBTS	Accelerator	2-mercaptobenzothiazole mercaptobenzothiazole zinc salt dibenzothiazyl monosulfide benzothiazole hydrogen sulfide aniline carbon disulfide phenyl mercaptan
MPTD	Accelerator	methylphenyldithiocarbamic acid- methylphenylammonium salt methylphenyldithiocarbamic acid zinc salt N-methylaniline carbon disulfide carbonyl sulfide sym-dimethyldiphenylthiourea N-methyldiphenylamine N,N-dimethylaniline methyl isothiocyanate phenyl isothiocyanate N-nitroso-N-methylamine 2-cyanoethyl methylphenyl dithiocarbamate
PAN	Antidegradant	1,2-naphthoquinone-1-anil
Styrenated diphenylamines	Antidegradant	monostyrenated diphenylamine distyrenated diphenylamine-N-oxide diphenylamine aniline styrene

4.5.3.3 Factors Affecting the Formation of the Breakdown Products

The data obtained showed that a complex situation exists, with the breakdown products formed in a given rubber compound being dependent on a number of factors, including:

1. The base rubber type – due to reactions with residual monomers in the synthetic rubbers and other substances such as formaldehyde in the case of natural rubber.

2. The other ingredients present – breakdown products of an individual curative will react with breakdown products from other curatives within the rubber, and with other ingredients such as the zinc oxide cure co-agent.

3. Concentration of ingredients – the range of breakdown products formed from a particular compound can depend upon the level of the ingredients present due to their solubility in the rubber.

4. The heat history – e.g., cure time/temperature and ageing conditions.

Therefore not all of the breakdown products that are predicted for a particular compound from the list of ingredients are present in all circumstances. The GC-MS fingerprint work on the rubber compounds identified between 25% and 50% of the predicted species. It was also thought likely that some products were not detected because they were below the detection limit (~1 ppm in the rubber) of the analytical method used.

To illustrate the approach used, the breakdown products predicted for one of the compounds produced during the course of this project, the nitrile compound 351N, are presented in **Table 12**. The breakdown product data obtained on this rubber by GC-MS and LC-MS is described in Section 4.5.3.4. The formulation of this nitrile rubber is shown below.

Nitrile rubber 351N (e.g., of the type used for milk liners)

Ingredients	phr
Breon N36C60	100
Zinc oxide	5
Stearic acid	2
HAF N330	15
Translink 77	15
DOP	5

Sulfur MC	1.5
DPG	0.15
MBTS	1.5
Rhenogran MPTD70	0.29
Rhenofit PAN	1
Wingstay 29	1.43

Notes

Breon N36C60 = Acrylonitrile-butadiene copolymer (Nitrile) rubber

HAF N330 = Carbon black filler

Translink 77 = Calcined and surface modified clay filler with vinyl functional surface modification

DOP = Dioctyl phthalate

DPG = Diphenyl guanidine (curative)

MBTS = Mercaptobenzothiazole disulfide (curative)

Rhenogran MPTD = Dimethyl diphenyl thiuram disulfide (curative)

Rhenofit PAN = N-phenyl-1-naphthylamine (antidegradant)

Wingstay 29 = Mixture of styrenated diphenylamines (antidegradant)

4.5.3.4 Fingerprinting of the Breakdown Products

Samples of the nitrile compound 351N described in Section 4.5.3.3 were submitted for analysis by both GC-MS and LC-MS to determine the low molecular weight species within it. The techniques used are described below and the resulting data was then used to assess the accuracy of the predictions shown in **Table 12**.

Gas Chromatography-Mass Spectrometry

Both of the two forms of GC-MS that are useful in studying breakdown products (i.e., headspace GC-MS and solvent extraction GC-MS) were used in the analysis of the nitrile rubber. In the first technique the rubber sample was heated to around 150 °C and the volatiles produced injected into the GC-MS via a cold trap. This technique has the advantages of being sensitive and there is no solvent to mask the early eluting species. With solvent extraction GC-MS, a solvent (acetone) is used to extract the rubber, and the extract is injected into the GC-MS. This approach has the advantage over the headspace technique in that semi-quantitative data or, in the case of a single compound, quantitative data, can be obtained (but some solvent selectivity can play a part). Its other advantage is that it is possible to study a higher molecular weight range.

The headspace GC-MS and solvent extract GC-MS chromatograms for the nitrile compound are shown in **Figures 9** and **10**.

The species identified in the chromatogram in **Figure 9** are shown in **Table 13**.

The species identified in **Figure 10** are shown in **Table 14**.

Liquid Chromatography-Mass Spectrometry (LC-MS)

LC-MS is an increasingly affordable technique, which provides the advantage over conventional HPLC instruments of having a universal detector (i.e., the mass spectrometer). A methanol extract of the nitrile rubber was injected into the instrument and components identified from their mass spectra, mainly by using the molecular ions. The advantage of this technique over GC-MS is that species that are thermally labile and/or of a high molecular weight (e.g., species in oligomeric antidegradants) can be identified. The disadvantage is that there are no commercial LC-MS libraries available at present and so there is not the same ease of identification that exists with GC-MS - where comprehensive 70 eV fragmentation voltage libraries have been available for many years.

The LC-MS chromatogram for the methanol extract of the nitrile rubber compound 351N is shown in **Figure 11**. The species identified are shown in **Table 15**.

It is apparent from the list that the majority of the species detected with this compound are the chemical additives in their unchanged, unreacted form. This indicates that LC-MS is less effective when used as a general screening technique than GC-MS which has better resolution and a lower detection limit for this type of work. Once again though, this work demonstrates that LC-MS has the advantages of being able to identify high molecular weight species (such as the oligomers present in antidegradants – e.g., Wingstay 29), and of not degrading thermally labile compounds (e.g., MBTS).

There follows a summary of the results of the GC-MS and LC-MS fingerprint work carried out on compound 351N.

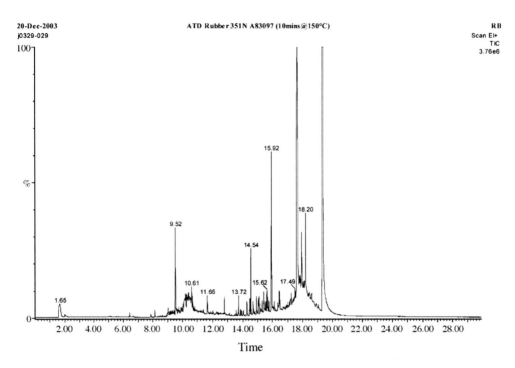

Figure 9. Headspace GC-MS chromatogram for the nitrile compound 351N

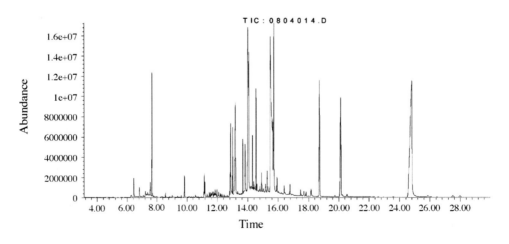

Figure 10. Solvent extract (acetone) GC-MS chromatogram for the nitrile
351N compound

Table 13. Assignments for the peaks present in Figure 9	
Retention Time (minutes)	**Assignment**
1.99	acetone
2.08	acrylonitrile
2.11	carbon disulfide
6.40	hexamethylcyclotrisiloxane
6.64	4-vinylcyclohexene
8.13	cycloalkane (C_9H_{18})
8.64	dodecene
9.04	aniline
9.52	*N*-methylaniline
10.08	acetophenone
10.30	α,α-dimethylbenzyl alcohol
10.20 - 10.67	dodecenes
10.61	1-octanethiol
11.65	benzothiazole
12.02	isopropylcumene or dimethyltetralin
12.79	alkane $C_{14}H_{30}$
13.72	2,6-*di-tert*-butyl-*p*-cresol (BHT)
13.84	alkylbenzene
14.54	diphenylamine
14.68 - 15.74	alkylbenzenes
15.92	dibutyl phthalate or monooctyl phthalate
17.64	*N*-phenyl-1-naphthylamine (phenyl-α-naphthylamine)
17.93	α-phthalate
18.20	monostyrenated diphenylamine*
19.36	bis(2-ethylhexyl) phthalate
* The nomenclature denotes an adduct of one molecule of styrene and one molecule of diphenylamine (molecular weight 273)	

Table 14. Assignments for the peaks present in Figure 10	
Retention Time (minutes)	**Assignment**
6.43	aniline
6.81	3-cyanocyclohexene
7.23	acetophenone + dodecene
7.56	2-methylindoline + dodecene
7.68	1-octanethiol
9.82	α-methylstyrene
11.15	2-methylbenzothiazole
11.67	nonylphenol
11.97	triallyl cyanurate
12.85	hexadecanoic acid
12.99	eicosane (internal marker)
13.16	mercaptobenzothiazole
13.67	di-*n*-octyl disulfide
13.81	octadecanoic acid
14.00	*n*-phenyl-1-naphthylamine (PAN)
14.30	butyl-2-ethylhexyl phthalate
14.54	monostyrenated diphenylamine
14.91	bis(2-ethylhexyl phthalate)
15.73	monostyrenated diphenylamine
15.92	diisoctyl phthalate
16.77	*n*-phenylpalmitamide
17.68	monostyrenated diphenylamine
17.83	monostyrenated diphenylamine
18.16	*n*-phenylstearamide
18.71	distyrenated diphenylamine
20.12	distyrenated diphenylamine
24.80	distyrenated diphenylamine

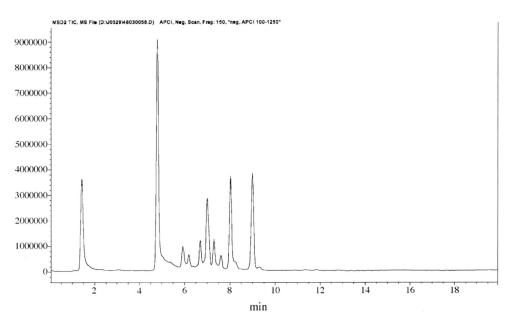

Figure 11. Solvent extract (methanol) LC-MS chromatogram for the nitrile compound 351N

Table 15. Assignments for the peaks present in Figure 11	
Retention Time (mins)	**Assignment**
1.44	MBTS** and/or MBT (breakdown product of MBTS)
3.95	DPG** (trace response)
4.78	phenyl-α-naphthylamine (PAN)**
5.89	styrenated diphenyl amine oligomer*
6.17	styrenated diphenyl amine oligomer*
6.68	unknown species
6.99	styrenated diphenyl amine oligomer*
7.29	styrenated diphenyl amine oligomer*
7.64	dioctyl phthalate**
8.03	styrenated diphenyl amine oligomer*
8.99	sytrenated diphenyl amine oligomer*
*Constituents of the Wingstay 29 oligomeric styrenated diphenylamine antioxidant – the three pairs are 1:1, 2:1 and 3:1 with respect to the stoichiometry of the styrene and diphenyl amine (with two isomers in each case) ** Original ingredients	

The data obtained on the nitrile compound by both GC-MS and LC-MS has revealed the presence of both the original ingredients (or impurities within them) and various breakdown products.

Examples of the original ingredients found include:

- MBTS

- Bis(2-ethylhexyl) phthalate

- *N*-phenyl-1-naphthylamine (PAN)

- Styrenated diphenylamines (adducts of 1:1, 2:1 and 3:1 stoichiometry)

Examples of impurities in the original ingredients found include:

- Acrylonitrile (residual monomer in polymer)

- Vinylcylohexene and cyanocyclohexene (polymerisation intermediates)

- Alkylbenzenes (possible impurities on carbon black)

- 1-Octanethiol/dioctyl disulfide (polymerisation residues - MW control agents)

Examples of the breakdown/reaction products found include:

- Aniline

- *N*-phenylpamitamide (aniline/$C_{15}CO_2H$ reaction)

- *N*-phenylstearamide (aniline/$C_{17}CO_2H$ reaction)

- Benzothiazole

- Carbon disulfide

- Diphenylamine

- Dodecenes

- 2-Mercaptobezothiazole

- *N*-Methylaniline

Not all of the theoretically predicted breakdown products were found. It may be that some others were present in amounts below the detection limits of these analyses or that the conditions for their formation were not favourable for this particular compound and its thermal history.

In other cases, species were seen that are recognised derivatives of some of the predicted breakdown products. One example is the 1:1 condensation product of an aromatic primary amine ($ArNH_2$) with acetone, i.e.,

$$ArNH_2 + O{=}CMe_2 \rightarrow ArN{=}CMe_2 + H_2O$$

The example here is isopropylidene aniline (acetone anil).

Some examples that fit none of the above criteria are minor species such as acetophenone, α,α-dimethylbenzyl alcohol (a recognised breakdown product of dicumyl peroxide). Dicumyl peroxide was not used in this rubber compound, but was used in other compounds manufactured during the course of this project and this is quite possibly a product of a small degree of cross-contamination.

4.5.3.5 Migration Behaviour of the Breakdown Products

Following the fingerprinting work to identify the breakdown products, migration work was carried out on ten of the nineteen rubber compounds. Worse case conditions were used with a range of fatty and aqueous food simulants to ensure that the maximum information on the migration potential of the breakdown products could be obtained. This approach was the most appropriate for this project, but it meant that higher migration values were obtained than would have been the case if the objective had been to obtain representative migration data on a specific rubber product using contact conditions that were appropriate for its service conditions. It was also the case that either aged or unaged examples of the compounds were chosen with a view to maximising the data that was produced, i.e., the unaged version was used if the ageing step had removed most of the breakdown products. Worse case conditions were also used in the migration tests that were carried out with the food products.

Contact Conditions used for the Migration Experiments – Food Simulants and Food Products

The conditions under which the rubbers contacted the food simulants are given in **Table 16.**

A brief description of the sample preparation method used to generate the migration data in the food simulant experiments is provided below.

In each case, a 10 cm × 5 cm sample of rubber was placed into a tube to which 100 ml of preheated simulant was added. After the contact period, the rubber samples were removed and the extract was cooled down to room temperature. It was apparent that some of the extracts were highly coloured after cooling down. The extracts were

Table 16. Conditions used for migration tests using food simulants		
Compound/condition	Test conditions	
	Time/temperature	Food simulant
20D aged	4 hours reflux 6 hours 60 °C	40% ethanol ethanol
239C unaged	4 hours reflux 6 hours 60 °C	15% ethanol ethanol
240C unaged	4 hours 100 °C 6 hours 60 °C	water ethanol
351N aged	4 hours 100 °C 6 hours 60 °C	water ethanol
504A aged	4 hours 100 °C 6 hours 60 °C	water ethanol
49V unaged	6 hours 60 °C	ethanol
242C unaged	4 hours 100 °C 4 hours reflux 6 hours 60 °C	3% acetic acid 5% ethanol ethanol
195S aged	6 hours 60 °C 4 hours 100 °C	ethanol 3% acetic acid
51B unaged	6 hours 60 °C	ethanol
201E aged	6 hours 60 °C	ethanol

then gently evaporated down to 10 ml using a rotary evaporator and analysed by GC-MS and/or LC-MS.

The ten rubber compounds were also contacted with real foodstuffs. A summary of the foodstuff/test condition combinations used for these compounds is given in **Table 17**.

Notes on the choice of food products

1. The main factor that influences the type of rubber to be used in a food contact situation is the service temperature – some rubbers having a better level of thermal stability than others (see Section 2.1 and **Table 3**). Given that a lot of use occurs at a relatively low temperature, it is therefore the case that a large range of rubber compounds can potentially be used in contact with any given food product. It was therefore decided to use one food product, olive oil for all the rubber compounds, to enable a degree of comparison to be made.

Table 17. Conditions used for migration tests using real foodstuffs		
Compound/condition	Test conditions	
	Time/temperature	Foodstuff
20D aged	6 hours 40 °C 4 hours 100 °C	whisky olive oil
239C unaged	6 hours 40 °C 4 hours 100 °C	beer (lager) olive oil
240C unaged	4 hours 100 °C	olive oil
351N aged	4 hours 100 °C	olive oil
504A aged	4 hours 100 °C	olive oil
49V unaged	4 hours 100 °C	olive oil
242C unaged	6 hours 40 °C 4 hours 100 °C	white wine olive oil
195S aged	6 hours 40 °C 4 hours 100 °C	apple juice olive oil
51B unaged	4 hours 100 °C	olive oil
201E aged	4 hours 100 °C	olive oil

2. For some of the rubber compounds (e.g., the EPDM ones), it was possible that they could also be used in contact with aqueous foodstuffs, e.g., whisky, beer, apple juice and white wine. In these cases a maximum temperature of 40 °C was used, as higher temperatures were not practically possible, i.e., higher temperatures might destroy or at least change the foodstuff significantly.

3. Some rubber compounds are of a type that would have an enhanced suitability for use with fatty foodstuffs, e.g., cream, butter, chocolate, milk, meat etc. This would normally relate to the levels and types of additives that are typically compounded into them. Consideration was given to using these actual products for the tests, but because of the well documented problems with the migration testing and subsequent analysis, it was decided to use the olive oil instead, with a time/temperature combination of 4 hours at 100 °C being chosen.

A brief description of the sample preparation methods used to generate the migration data in the food contact experiments is provided below.

A strip (10 cm × 5 cm) of rubber (= 0.5 dm^2 × 2 = 1 dm^2) was placed in a tube and 100 ml of foodstuff was added. At the end of the migration experiment the rubber sample was removed and the solution was cooled down to room temperature. The olive oil extracts were then transferred to a separation funnel, 50 ml of pentane was added and this mixture was extracted with 100 ml of ethanol for 2 minutes. The ethanol-layer was centrifuged and subsequently analysed by GC-MS.

In the case of the aqueous foodstuffs, the migration extracts were transferred to a separation funnel and were extracted with 100 ml of a mixture of pentane/diethylether 2:1 *v/v*. The pentane/diethylether fraction was dried using Na$_2$SO$_4$ and subsequently analysed by GC-MS. The whisky solution was diluted with water prior to extraction with pentane/diethylether 2:1 *v/v* to decrease the amount of ethanol and thus increase the extraction efficiency.

The foodstuff extracts were analysed by GC-MS with a view to finding specific components that had also been observed in the GC-MS chromatograms of ethanol food simulant migration extracts (see **Table 16** and **Figures 12 to 14**).

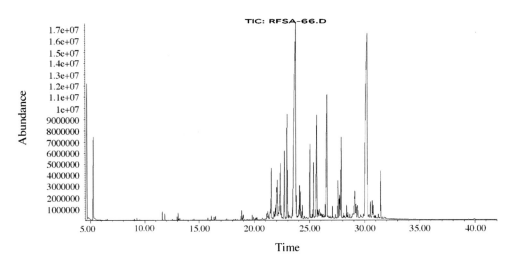

Figure 12. GC-MS chromatogram of the migrants from 195S detected in 100% ethanol

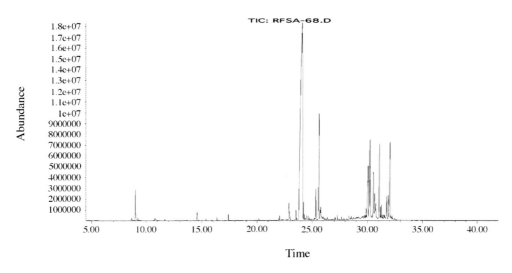

Figure 13. GC-MS chromatogram of the migrants from 201E detected in 100% ethanol

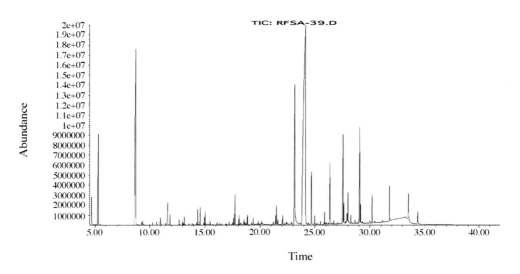

Figure 14. GC-MS chromatogram of the migrants from Compound 242C detected in 100% ethanol

Analysis Conditions used for the Migration Samples – Food Simulant and Food Products

The food simulant and food migration samples prepared as described above were analysed by GC-MS and LC-MS using the following conditions:

Gas Chromatography-Mass Spectrometry

1. Food simulants

   Column: J&W 19091S-433, HP-5MS (30 m × 250 μm × 0.25 μm)
   Injection: splitless, 1 μl
   Temperature program: 10 °C → 320 °C at 10 °C/min

2. Food products

   . Extracted aqueous foodstuffs:
   Column: J&W 122-5533, DB-5MS (30 m × 250 μm × 1.0 μm)
   Injection: splitless, 1 μl
   Temperature program: 10 °C → 320 °C at 10 °C/min

   Extracted olive oil:
   Column: J&W 19091S-433, HP-5MS (30 m × 250 μm × 0.25 μm)
   Injection: splitless, 1 μl
   Temperature program: 10 °C → 320 °C at 10 °C/min

Liquid Chromatography-Mass Spectroscopy (only applied to the food simulants)

ThermoFinnigan LCQ classic
   Column: Xterra RP18 (150 mm × 3.0 mm, 3.5 μm)
   Flow: 0.4 ml/min, split to MS → 50 μl/min
   Injection volume: 10 μl
   Gradient: 99% 10 mM NH_4Ac (pH 5.5) /1% MeCN → 1% NH_4Ac (pH 5.5) /99% MeCN in 30 min
   Total runtime: 50 min
   Ionisation: APCI in positive and negative mode, *m/z* 50-1200
   Tuning: BHT (negative mode), mercaptobenzothiazole (positive mode)
   Blank solvents and standard solutions of BHT, benzothiazole, mercaptobenzothiazole, N-phenyl-1-naphthylamine and cyclohexylphthalate were injected for reference.

GC-MS Data Obtained for the Food Simulants

In order to illustrate the information that was obtained during this stage of the project, the data obtained by GC-MS on food simulants that had contacted three of the rubber compounds (195S, 201E and 242C) is provided in this section.

Figure 12 and **Table 18** show the migrants from compound 195S.

The data obtained on Compound 201E is shown in **Figure 13** and **Table 19**.

The data obtained on Compound 242C is shown in **Figure 14** and **Table 20**.

LC-MS Data Obtained for the Food Simulants

The data obtained by LC-MS on food simulants that had contacted three of the rubber compounds (195S, 201E and 242C) is provided in this section.

The LC-MS chromatograms for the 3% acetic acid that had contacted Compound 195S are shown in **Figures 15** and **16**.

A few of the species present in **Figures 15** and **16** could be identified. These are listed in **Table 21**.

The LC-MS chromatograms for the 100% ethanol that had contacted Compound 201E are shown in **Figures 17** and **18**.

Although there is a large number of species peaks present in both **Figures 17** and **18** none of them could be identified despite searching the data using the predicted breakdown product list and the list of starting ingredients. This illustrates the problems that can exist with LC-MS work, where there are no commercial libraries available.

The LC-MS chromatograms for the 3% acetic acid and 15% ethanol that had contacted Compound 242C are shown in **Figures 19** to **22**.

Unfortunately, no species could be identified in the positive or negative APCI chromatograms for the 3% acetic acid simulant, and only one peak could be identified for the 15% ethanol sample – the antidegradant additive *N*-phenyl-1-naphthylamine in the positive APCI mode.

Table 18. Migrants detected by GC-MS - Compound 195S in contact with 100% ethanol (RT is retention time)				
Food Simulant : 100% ethanol		Peak Area	Amount (µg/ml)	Amount (µg/6 dm²)
Migration conditions: 6 hour at 60 °C				
Area sample 1 dm²				
RT (min)	Component			
4.66	piperidine	232946050	232.9	13977
5.27	morpholine	235495177	235.5	14130
6.54	5-methyl-2-hexanone	1695846	1.7	102
8.99	aniline	2363787	2.4	142
11.84	1-piperidinecarboxaldehyde	10107670	10.1	606
13.06	1-piperidinecarboxylic acid ethyl ester	9514780	9.5	571
13.12	2-phenoxy ethanol	1997836	2.0	120
13.22	benzothiazole	2932090	2.9	176
14.55	*m/z* 29, 42, 55, 69, 84, 132	2088552	2.1	125
15.81	octahydro-dimethyl-7-(1-methylethenyl) naphthalene	3141726	3.1	189
16.50	1H-isoindole-1,3-dione	12551022	12.6	753
17.49	*m/z* 51, 78, 106, 164	2235453	2.2	134
18.43	1,1-diphenyl-2-(2,4,6-trinitrophenyl) hydrazine	1265755	1.3	76
18.76	1,1'-carbonylbis piperidine	5319000	5.3	319
18.99	2-benzothiazolone	6866924	6.9	412
19.90	tetradecanoic acid	5773271	5.8	346
20.23	N-butyl benzenesulfonamide	3101775	3.1	186
21.99	N-phenyl-1,4-benzenediamine	95905939	95.9	5754
23.74	N-(1-methylethyl)-N'-phenyl benzenediamine	1843223887	1843.2	110593
24.10	octadecanoic acid, ethyl ester	10357369	10.4	621
25.55	octahydro-dimethyl-7-(1-methylethenyl) 1-phenanthrenecarboxylic acid	17640883	17.6	1058
27.05	DOP	16714678	16.7	1003

Table 19. Migrants detected by GC-MS - Compound 201E in contact with 100% ethanol				
Food Simulant: 100% ethanol		Peak Area	Amount (μg/ml)	Amount (μg/6 dm²)
Migration conditions: 6 hours at 60 °C				
Area sample: 1 dm²				
Rt (min)	Component			
5.29	morpholine	1768695	1.8	106
8.67	isocyanaoto benzene	10849237	10.8	651
9.01	aniline	72069911	72.1	4324
10.73	2-methyl benzenamine	2096406	2.1	126
10.78	amino toluene	12335040	12.3	740
11.66	chloroaniline	2348950	2.3	141
13.58	caprolactam	1165354	1.2	70
14.59	*N*-phenyl formamide	24584621	24.6	1475
15.37	*N*-phenyl acetamide	3104499	3.1	186
15.91	carbamic acid, phenyl, ethyl ester	1232235	1.2	74
17.40	dodecanethiol	8573855	8.6	514
20.15	*N,N'*-methanetetraylbis benzenamine	2602873	2.6	156
20.21	*N*-butyl benzenesulfonamide	1452026	1.5	87
21.96	*N*-phenyl-1,4-benzenediamine	4787719	4.8	287
22.94	sulfur	12486089	12.5	749
22.96	phenothiazine	2210000	2.2	133
23.59	pyrene	1816054	1.8	109
24.50	*N,N'*-diphenyl urea	7014259	7.0	421
24.63	*N*-phenyl-2-naphthalenamine	5338846	5.3	320
27.04	DOP	3357762	3.4	201

Table 20. Migrants detected by GC-MS - Compound 242C in contact with 100% ethanol				
Food Simulant: 100% ethanol		Peak Area	Amount (µg/ml)	Amount (µg/6 dm²)
Migration conditions: 6 hours at 60 °C				
Area sample: 1 dm²				
Rt (min)	Component			
4.65	piperidine	52385307	52.4	3143
5.03	*N,N*-dimethyl formamide	1686649	1.7	101
5.29	morpholine	313293438	313.3	18798
7.17	carbamic acid, dimethyl ethyl ester	1326456	1.3	80
8.73	*N*-butyl-1-butanamine	838304229	838.3	50298
9.23	tetramethyl urea	2992342	3.0	180
9.36	*N,N*-dimethyl-2-butoxyethylamine	6445774	6.4	387
9.41	1-isothiocyanato butane	3289303	3.3	197
9.99	*N*-butyl formamide	1732179	1.7	104
10.63	acetophenone	5347446	5.3	321
10.97	dimethyl benzenemethanol	18334113	18.3	1100
11.86	1-piperidinecarboxaldehyde	17335441	17.3	1040
12.68	4-acetyl morpholine	8626811	8.6	518
13.07	1-piperidinecarboxylic acid ethyl ester	2854133	2.9	171
13.14	tetramethyl thiourea	12831964	12.8	770
14.34	*N,N*-dimethyl formamide	24397774	24.4	1464
15.29	trimethyl thiourea	2599292	2.6	156
18.76	1,1-carbonylbis piperidine	1820574	1.8	109
19.21	heptadecane	3055309	3.1	183
19.80	alkyl piperidine	1195665	1.2	72
22.23	hexadecanoic acid, ethyl ester	2049575	2.0	123
23.17	octadecanol	485218208	485.2	29113
23.24	heneicosane	3253000	3.3	195
23.69	alkyl alcohol	2428779	2.4	146
24.15	*N*-phenyl-1-naphthalenamine	2585400637	2585.4	155124
24.97	alkene	12680306	12.7	761

Table 20. *Continued*				
25.86	hexanedioc acid didodecyl ester	19959809	20.0	1198
25.97	hexenedioc acid didodecyl ester	2620788	2.6	157
26.70	hexanedioc acid ditridecyl ester	5914660	5.9	355
27.04	DOP	2638315	2.6	158
28.02	Ziram	83546328	83.5	5013
28.08	hexanedioc acid dialkyl ester	1851175	1.9	111
28.28	hexanedioc acid dipentadecyl ester	11742999	11.7	705
28.40	hexenedioc acid dipentadecyl ester	1742827	1.7	105
29.08	hexanedioc acid dihexadecyl ester	255836522	255.8	15350
29.16	hexenedioc acid dihexadecyl ester	30794539	30.8	1848
33.54	zinc bis(*N,N*-dihexyldithiocarbamate)	89619321	89.6	5377
34.37	Irganox 1076	34927366	34.9	2096

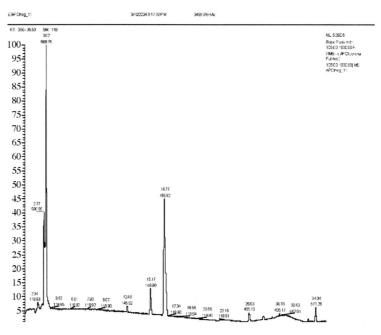

Figure 15. LC-MS chromatogram of Compound 195S, 4 hours @ 100 °C, 3% acetic acid (LC-MS in negative APCI mode)

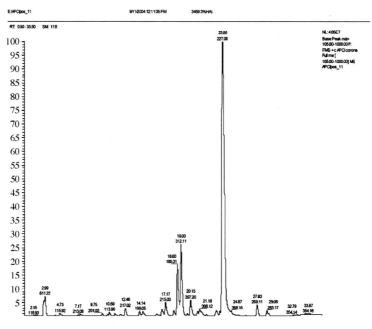

Figure 16. LC-MS chromatogram of Compound 195S, 4 hours @ 100 °C, 3% acetic acid (LC-MS in positive APCI mode)

Table 21. Migrants identified by LC-MS - Compound 195S in contact with 3% acetic acid			
	3% acetic acid migration extract (4 hours 100 °C)		
	TR (min)	**Mass**	**Component**
Positive mode	18.6	185.0	*N*-phenyl-1,4-phenylenediamine
	20.2	397.2	2-mercaptobenzothiazole zinc salt
	23.8	227.1	*N*-isopropyl-*N'*-phenyl-*p*-phenylenediamine
Negative mode	16.8	165.9	2-mercaptobenzothiazole

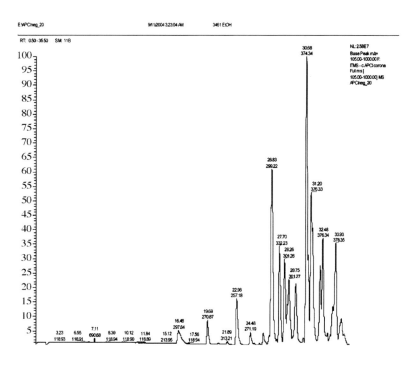

Figure 17. LC-MS chromatogram of Compound 201E, 6 hours @ 60 °C, 100% ethanol (LC-MS in negative APCI mode)

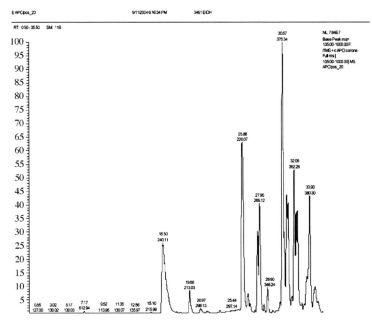

Figure 18. LC-MS chromatogram of Compound 201E, 6 hours @ 60 °C, 100% ethanol (LC-MS in positive APCI mode)

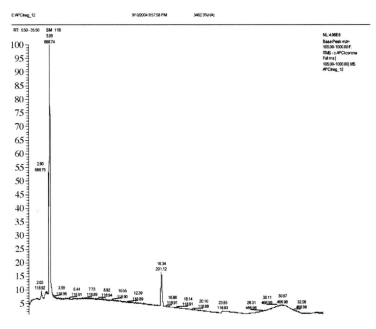

Figure 19. LC-MS chromatogram of Compound 242C, 4 hours @ 100 °C, 3% acetic acid (LC-MS in negative APCI mode)

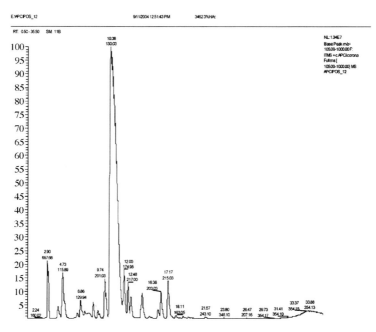

Figure 20. LC-MS chromatogram of Compound 242C, 4 hours @ 100 °C, 3% acetic acid (LC-MS in positive APCI mode)

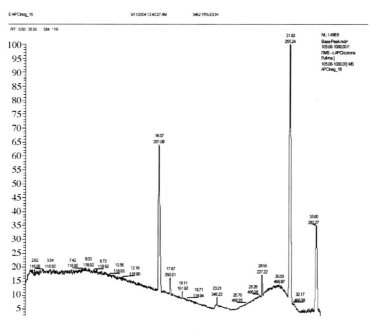

Figure 21. LC-MS chromatogram of Compound 242C, 4 hours reflux, 15% ethanol (LC-MS in negative APCI mode)

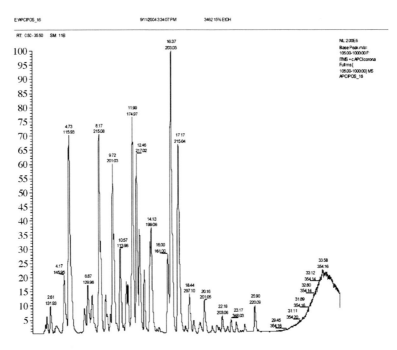

Figure 22. LC-MS chromatogram of Compound 242C, 4 hours reflux, 15% ethanol (LC-MS in positive APCI mode)

GC-MS Data Obtained on the Food Products

The food products that were contacted with the rubber samples were analysed by GC-MS to identify any migrants that were present. The difficulties that can be encountered in the preparation of the samples to ensure compatibility, the fact that the majority of the species were of a low molecular weight, and the problems that had been experienced in identifying peaks present in the food simulant data (e.g., **Figures 17** to **20**), meant that LC-MS was not used on the food products in this project.

A summary of the most abundant species that were found in the food products by GC-MS is given in **Table 22**.

As expected, the experiments carried out in olive oil identified the same components that were present as large peaks in the migration tests carried out using the 100% ethanol fatty food simulant (**Tables 18** to **20**). However, due to the different sample work-up and time/temperature conditions, the concentration of these components was different.

Compound/ Condition	Foodstuff	Component
Table 22. Most abundant species that were found to migrate from the test rubber compounds into the foodstuffs by GC-MS analysis		
20D aged**	Whisky	See discussion below
	Olive oil	4,4'-thiobis[2-(1,1-dimethylethyl)-5-methyl-phenol
239C unaged**	Beer	Peak 1: 4-oxopentanoic acid, *n*-butyl ester
		Peak 2: Glutaric acid, monobutyl ester
	Olive oil	Peak 3: Triallyl cyanurate 4-oxopentanoic acid, *n*-butyl ester
		Glutaric acid, monobutyl ester
		Triallyl cyanurate
240C unaged	Olive oil	9,10-dihydro-9,9-dimethyl-acridine
		N-phenyl-1-naphthylamine
351N aged	Olive oil	*N*-phenyl-1-naphthylamine
		bis(2-ethylhexyl)phthalate (DOP)
504A aged	Olive oil	*N*-isopropyl-*N'*-phenyl-*p*-phenylenediamine
		4,4'-thiobis[2-(1,1-dimethylethyl)-5-methyl-phenol
49V unaged	Olive oil	Unknown*
		Unknown*
		Unknown*
242C unaged	White wine	See discussion below
	Olive oil	*N*-butyl-1-butanamine
		N-phenyl-1-naphthylamine
195S aged	Apple juice	Peak 1 : *N*-isopropyl-*N'*-phenyl-*p*-phenylenediamine
	Olive oil	*N*-isopropyl-*N'*-phenyl-*p*-phenylenediamine
51B unaged	Olive oil	Aniline
		N-phenyl-octadecanamide
201E aged	Olive oil	*N*-phenyl-1-naphthylamine

*No match found in the GC-MS library

**Aged or unaged versions used as considered most appropriate (see Section 4.5.3.5)

Peak 1, 2, 3 etc = Peaks designated in GC-MS chromatograms (see Figure 23)

For the aqueous foodstuffs, the results were less straightforward. For apple juice and beer, components were found that were also observed for olive oil and ethanol. However, for whisky and white wine, no components could be observed that were also present in olive oil or the ethanol migration samples. This might be due to several reasons, for example the chemical structure of the components and, hence their solubility in the two types of media could be completely different and there could be hydrolysis reactions going on in the case of the aqueous foods, which would result in a different set of migrants.

The GC-MS chromatograms that were obtained with the beer, whisky, apple juice and white wine migration samples are shown in **Figure 23**. The peaks designated in the chromatograms (Peak 1 etc.) are shown in **Table 22**.

The GC-MS chromatograms that were obtained with the olive oil migration samples (together with the olive oil blank) are shown in **Figure 24**. The peaks arrowed are the rubber migrants that are assigned in **Table 22**. It can be seen that the majority of the peaks in the samples are also present in the olive oil blank.

Work using GC-MS on foods was complicated by the need for an intermediate extraction step. Nevertheless, it was possible to detect breakdown products in some cases (e.g., peroxide curative products from a peroxide cured EPDM (Compound 239C) migrated into beer and olive oil). Few migrants were detected overall even though worst case conditions were used.

Difficulties were encountered in identifying substances in the aqueous food simulants and food products and this was thought to be due, at least in part, to the generation of a secondary set of breakdown products by hydrolysis reactions.

4.5.3.6 Overall Summary of the Migration Data

Breakdown Product Data Obtained Using Food Simulants

Migration experiments were carried out using ten of the nineteen rubber compounds specially prepared for this project (see Section 4.5.3.1) and a range of food simulants. Worst case conditions, in terms of temperature and time, were used in all cases and this resulted in some relatively high migration values. This is particularly true with the fatty food simulant (100% ethanol in this case instead of 95% ethanol) where additives such as the plasticiser DOP have been found to migrate from compounds such as 351N – a semi-quantitative result of 35 mg/6 dm^2 being obtained by GC-MS analysis. It has also been possible to detect by GC-MS a large number

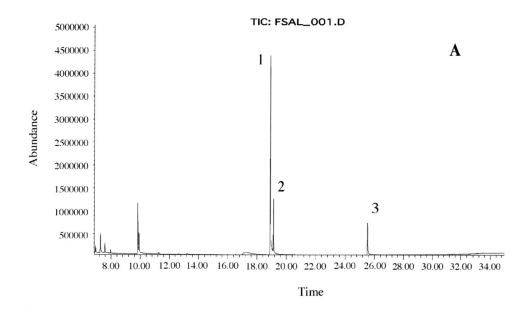

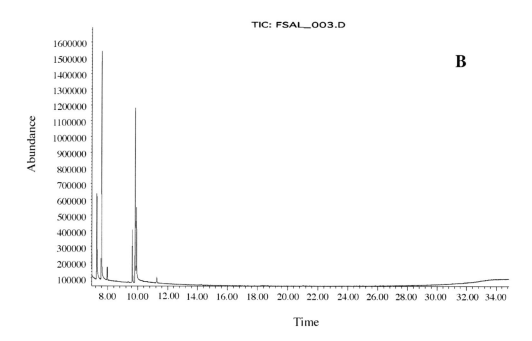

Figure 23 A-B. Examples of GC-MS chromatograms of migrants from the rubber samples into real foodstuffs. Peaks indicated are assigned in Table 22

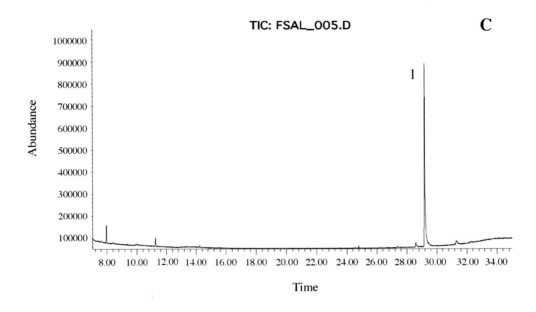

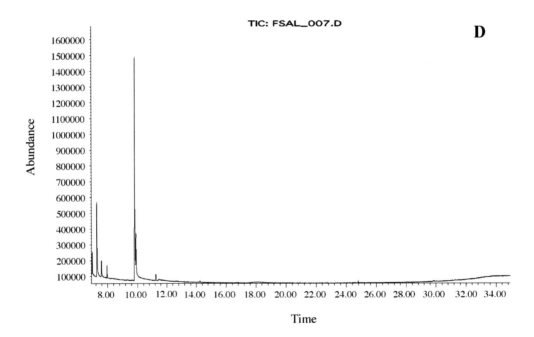

Figure 23 C-D. Examples of GC-MS chromatograms of migrants from the rubber samples into real foodstuffs. Peaks indicated are assigned in Table 22

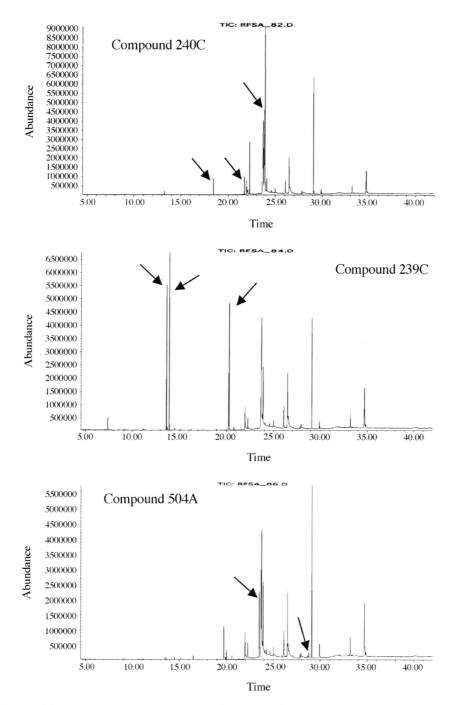

Figure 24. GC-MS chromatograms of extracted olive oil migration samples of the various rubber samples. Peaks indicated are those assigned (mostly as unknown) in Table 22

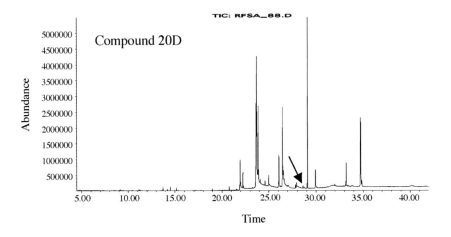

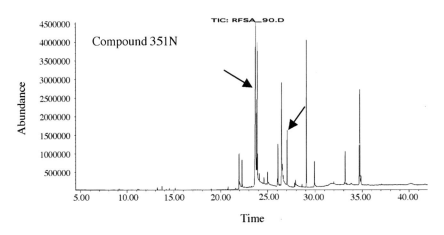

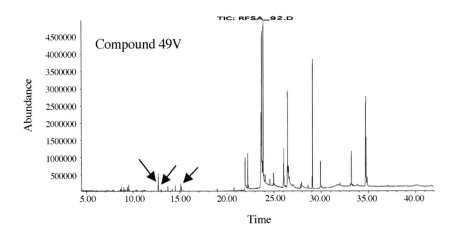

Figure 24. *Continued*

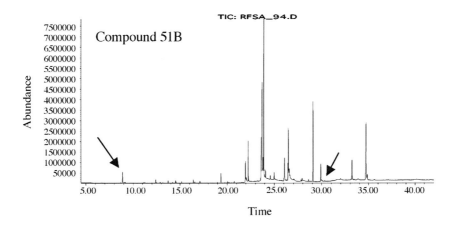

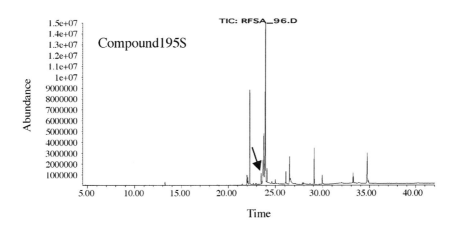

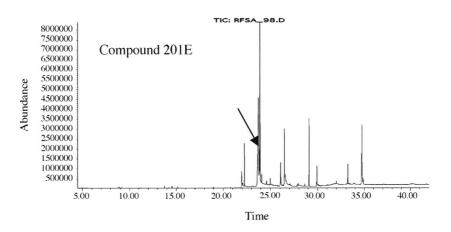

Figure 24. *Continued*

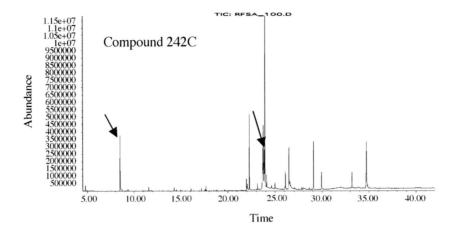

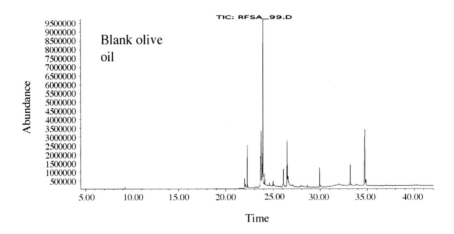

Figure 24. *Continued*

of migrants in the ethanol – over a hundred in a number of cases (e.g., 195S and 242C). Unfortunately, the complexity of rubber matrices was underlined by the fact that it was only possible to identify relatively few of these using the mass spectral libraries. Of the specific migrants that were identified, a large number were either original compounding ingredients, particularly plasticisers and antidegradants, or breakdown products. For each of the rubber compounds, it was possible to detect in the food simulants a number of the predicted breakdown products. This is illustrated in **Table 23** by reference to the GC-MS data obtained using 100% ethanol with Compound 201E.

Table 23. Table showing how many predicted breakdown products for Compound 201E were detected in 100% ethanol		
Curative/ antidegradant	Predicted breakdown product	Detected in 100% ethanol
Accelerator Thio No. 1	N,N'-diphenylthiourea (Original Compound)	No
	aniline	Yes
	phenyl isothiocyanate	Yes
	N,N'-diphenylurea	Yes
	phenyl isocyanate	Yes
	diphenylamine	Yes
	sulfur dioxide	No
Sulfur	sulfur (original compound)	Yes
	sulfur dioxide	No
Vanax PML	N,N'-di-o-tolylguanidine salt of dicatechol borate (original compound)	No
	N,N'-di-o-tolylguanidine	No
	dicatechol borate	No
	catechol	No
	N,N'-di-o-tolylthiourea	No
	o-toluidine	Yes
	tris(o-tolylamino)-1,3,5-triazine	No
	di-o-tolylcarbodiimide	No
	ammonia	No
	o-tolyl isothiocyanate	No
	N,N'-di-o-tolylurea	Yes
	o-tolyl isocyanate	No
Rhenofit PAN	N-phenyl-1-naphthylamine (original compound)	Yes
	1,2-naphthoquinone-1-anil	No
Santoflex 13	N-(1,3-dimethybutyl)-N'-phenyl-p-phenylenediamine (original compound)	Yes
	aniline	Yes
	diphenylamine	Yes
	N-phenyl-1,4-phenylenediamine	Yes
	4-methylpentan-2-one	No
	N-(1,3-dimethybutyl)-N'-phenyl-1,4-benzoquinone diimine	No

Where compounds have not been detected, this is due to a number of reasons. For example:

i)   some are only minor breakdown products and so may be present at levels below the detection limit of the method,

ii) some are low molecular weight compounds (e.g., the gases) that may have been lost from the simulant during the test period; or

iii) their mass spectra may not be present on the database which makes their detection difficult without running standards.

With respect to the LC-MS work, substances were detected in all the simulant/compound combinations. The number of substances varied, ranging from a few (e.g., 20D and 49V) to greater than twenty (e.g., 242C and 351N). Unfortunately, the lack of availability of commercial LC mass spectra databases has severely restricted the number of substances that could be identified. Those substances that were identified were mainly original compounding ingredients in the rubbers, particularly antidegradants, and these have been identified based on their molecular masses. By the use of molecular weight data, it was possible to detect some breakdown products for the test compounds. It was not possible to use a single calibrant approach (as in the case of the GC-MS data) to obtain semi-quantification data on these substances due to the more unpredictable ionisation of the analytes and the response of the mass spectrometer detector.

There are thought to be several reasons why so many of the substances in the LC-MS data could not be identified by reference to the predictive breakdown product lists. For example:

i)   in the aqueous simulants, hydrolysis of the breakdown products could occur which would produce a secondary series of products, and

ii) adduct ions can be formed with the mobile phase which changes the molecular weight of the target species.

Breakdown Product Data Obtained Using Food Products

GC-MS was used to identify migrants in a range of food products. In all cases, the migration experiment was conducted and then an extraction step carried out on the food product using a GC compatible solvent. This work was not carried out quantitatively as the methods had not been validated, e.g., the extraction efficiency of the analytes from the food products was unknown. Work during this part of the project was also confined to specifically searching for major substances that had been detected initially in the food simulants.

It can be seen from **Figure 24** that, in the case of olive oil, the majority of the peaks present in the sample chromatograms are due to components extracted from the olive oil during the sample work up. Relatively few sample related substances were detected (these are indicated by arrows in **Figure 24**) and the majority were original compounding ingredients, particularly antidegradants. There were four instances where breakdown products were found:

1) 240C in contact with olive oil - 9,10-dihydro-9,9-dimethyl-acridine from the PAN antioxidant.

2) 242C in contact with olive oil - N-butyl-1-butanamine from the ZDBC accelerator.

3) 239C in contact with both beer and olive oil - 4-oxopentanoic acid, *n*-butyl ester, glutaric acid and monobutyl ester. Breakdown products of the peroxide curative were detected in both food products.

4) 51B in contact with olive oil – aniline from the DPG accelerator.

4.5.3.7 Overall Conclusions

The results obtained during the course of this major project demonstrated the importance of considering reaction/breakdown products as potential migrants from food contact elastomers. It is clear from the results obtained that starting substances alone should not be considered as the only potential migrants when evaluating these materials.

The project also demonstrated that in order to produce meaningful quantitative migration data on the breakdown products, work would have to be carried out on specific rubber products using contact conditions and simulants/foods that were appropriate for their service conditions.

4.5.4 Project A03046 - Silicones

4.5.4.1 Introduction

The aim of this research project entitled '*Chemical migration from silicones used in connection with food contact materials and articles*' was to provide detailed information on the types and composition of silicone based products used in contact with food, and to identify the extent to which migration of specific constituents into food could occur. As such the scope of the project included all silicone based food contact products, i.e., silicone fluids and emulsions, silicone pastes, silicone resins, silicone rubbers.

Only the information and data obtained on silicone rubbers will be discussed here.

The project ran from 2004 until 2005 and was divided into two main sections:

1) The initial phase of the project included a review of the Rapra in-house literature and, assisted by the Rapra Abstracts database, of the published literature on the types of silicone products used in the food industry and any available migration data. The aim was also to produce a predictive list of any potential migrants in them from the manufacturing, curing and compositional data.

2) The second stage involved obtaining commercial examples of food contact silicone rubbers and carrying out compositional analysis work using techniques such as GC-MS and LC-MS. Once this had been completed and the data obtained compared to the theoretical predictions produced in Stage 1, migration experiments were conducted using both food simulants and food products. Both overall migration and specific migration data was generated by these experiments.

4.5.4.2 Potential Migrants in Silicone Rubbers – Stage 1 of the Project

The literature review obtained information on the species within silicone rubbers that have the potential to migrate into food. The main groups of migrants can be summarised as follows:

1. Low molecular weight siloxane compounds

A prime example of the low molecular weight siloxanes (other than silicone oligomers) that can reside in silicone rubbers is the polymerisation short stop agent, trimethyl silanol. This compound is used to terminate a growing chain and hence control molecular weight. Although a high proportion of this species will therefore react and be bound into the matrix, enough remains free to be detected in most rubbers and, given its low molecular weight, it has the potential to migrate into food.

2. Silicone oligomers

As silicone rubbers are relatively simple materials containing a limited number of additives and modifiers, this class of species represents the most important potential migrants.

There are a number of types of silicone oligomers that can be present in silicone rubbers, including:

- Trimethylsiloxy-terminated polydimethylsiloxanes, MDnM

- Cyclic polydimethylsiloxanes, Dn
- Silanol-terminated linear polydimethylsiloxanes, HODnH
- Trimethylsiloxy/silanol-terminated polydimethylsiloxanes, MDnH

Each of the these polydimethyl siloxanes will provide a homologous series of compounds, distinguishable by characteristic molecular weights.

Working on the most abundant isotopes of the atoms present in the compounds, hexamethyldisiloxane (MM) has a principal molecular weight of 162.09, whilst the basic unit, D, has a molecular weight of 74.04.

a) The series MDnM will have a series of molecular weights given by:

162.09 + 74.04n   where n is zero or an integer

The above rearranges to: 14.01 + 74.04x   where x = n +2

The lower molecular weights of this series will have the following molecular weights:

162.1, 236.1, 310.2, 384.2, 458.25, 532.3, 606.3 etc.

b) The series, Dn, will have molecular weights given by:

74.04n   where n is an integer (3 or above)

The members of this series will have the following molecular weights:

222.1, 296.2, 370.2, 444.2, 518.3, 592.3, 666.4 etc

c) The series, HODnH, will have molecular weights given by:

18.01 + 74.04n   where n is an integer

It is unlikely that low molecular weight species of this type would be present as they would be unstable to hydrolytic condensation. If present they would be described by the following series:

92.1, 166.1, 240.1, 314.2, 388.2, 462.3, 536.3, 610.3 etc

It is possible from the above to summarise the series of oligomers that can be present in silicone rubbers as follows:

(i)   Cyclic oligomers – molecular weight = 74n

(ii) Methyl only terminated linear oligomers – molecular weight = 74.04n + 14.01

(iii)   Silanol only terminated linear oligomers – molecular weight = 74.04n + 18.01

(iv)   Silanol/methyl* terminated linear oligomers – molecular weight = 74.04n + 16.03

One of each end group type

where n = number of repeat units in the chain.

In addition, if there are any vinyl groups (C_2H_3) in the polymer backbone, which is often the case with silicone rubbers, the molecular weight of the oligomers will be increased by 12.00 for each vinyl unit.

If there are any phenyl groups (C_6H_5) in the polymer backbone, again a possibility with silicone rubbers, the molecular weight of the oligomers will be increased by 62.02 for each unit.

3. Cure system species

Peroxides are used extensively to cure silicone rubbers. These peroxides react during the curing process and leave breakdown products within the rubbers. The influence that these breakdown products have on the final product is greatly reduced by the fact that it is common practice to give silicone rubbers a postcure in an oven after the initial vulcanisation step. This postcure is important as it improves the cure state (which improves physical properties) and in removing the breakdown products it makes the product less susceptible to hydrolysis in service. It is therefore the case that the concentration of peroxide breakdown products remaining in silicone rubbers is very low and in fact it can be difficult to detect them using conventional analysis (e.g., GC-MS) techniques.

Examples of the major breakdown products that can be produced from two of the most commonly used peroxides in silicone rubbers are:

• Dicumyl peroxide - acetophenone and phenyl-2-propanol

• Dichlorobenzoyl peroxide - dichlorobenzoic acid and dichlorobenzene

The other major cure system that is used in food contact silicone rubbers employs metal catalysts.

The majority of these are based on platinum, with the use of compounds such as chloroplatinic acid. This is an area of active research and the precise platinum compound used by particular companies is a closely guarded secret. Fortunately, the metal itself is easily detectable in food and food simulants by established techniques such as inductively coupled plasma spectrometry (ICP) and limits for platinum are given in the German BfR Recommendation XV.

4. Oxidation products

Silicone rubbers are both manufactured (i.e., processed and cured), and used in service, at high temperatures. Although they are sufficiently stable to high temperatures (e.g., 200 °C) over relatively long periods of time (e.g., weeks) and retain good physical properties, chemical changes can occur within the materials if this period is extended to a number of months or years. There are two mechanisms which can generate low molecular weight species having the potential to migrate into food.

a) Main chain scission can occur leading to cyclisation and the formation of relatively low molecular weight cyclic oligomers

b) Oxidation of the alkyl groups attached to the silicone atoms can lead to the formation of oxidation products such as aldehydes (e.g., formaldehyde from methyl groups)

4.5.4.3 Data Obtained on Commercial Silicone Rubber Products – Stage 2 of the Project

Four commercially available food contact silicone rubbers were obtained for this stage of the project. They were:

1) One 60 IRHD hardness gum rubber cured using dicumyl peroxide

2) Two different grades of 60 IRHD hardness gum rubbers cured using a platinum catalyst

3) One 60 IRHD hardness liquid silicone rubber cured using a platinum catalyst

The low molecular weight species in these products were fingerprinted and then migration data was obtained using a range of food simulants and food products. To illustrate the data obtained during the course of this project, the results produced using two of the rubbers, the peroxide cured gum rubber and the platinum cured liquid rubber are described. These two have been chosen as they are illustrative of the range and spread of data obtained.

1. Fingerprinting work on the two rubbers to identify potential migrants

The fingerprinting data obtained on the two silicone rubbers can be summarised as follows:

a) **Cure system species**

(i) No evidence for the presence of any residual peroxide breakdown products in the peroxide cured rubber

(ii) In the case of the platinum cured rubber, 5.0 mg/kg of platinum was detected in the material

b) **Elements present**

Only silicone, carbon, hydrogen and oxygen were found at levels above 10 μg/g (the detection limit of the analytical method) in the case of the peroxide cured rubber.

For the platinum cured liquid rubber, several elements were found (**Table 25**).

c) Siloxane compounds present in both rubbers

In both cases the following types of siloxanes were detected by GC-MS:

• Trimethyl silanol (short stopper)

• Cyclic oligomers (n=3 to n>20): A positive identification of the n=3 to 8 oligomers was possible but the higher examples were outside the scanning range of the mass spectrometer (25 to 650 amu) and so the characteristic fragmentions were lost.

• Linear oligomers from n~5 to n>20

Individual masses of oligomers in an acetone extract ranged from 3 μg/g to150 μg/g with the cyclic hexamer being the most abundant.

The total siloxane oligomer content, from the GC-MS analysis of an acetone extract, for each rubber is shown in **Table 26**.

2. **Migration data obtained on the two rubbers using food simulants**

a) **Overall Migration Results**

The rubbers were contacted with the following simulants:

Distilled water	4 hours reflux
95% Ethanol	4 hours at 60 °C

The results obtained are shown in **Table 27**.

Table 25. Elements detected in the platinum cured liquid rubber in the fingerprinting work	
Element	Semi-quantitative concentration (mg/kg)
Aluminium	13
Calcium	86
Magnesium	53

Table 26. Total mass of siloxanes present in the two silicone samples	
Rubber sample	Total mass of oligomers (µg/g)
Platinum cured liquid rubber	981
Peroxide cured gum rubber	829

Table 27. Overall migration data obtained for the two silicone rubbers		
Rubber sample	Overall migration (mg/dm^2)	
	Distilled water	95% ethanol
Peroxide cured gum	5	29
Platinum cured liquid rubber	2	21

The values in **Table 27** can be compared to the Silicone Resolution limit of 10 mg/dm^2. The aqueous results are well within this value and, although the fatty food simulant (95% ethanol) data is in excess of it, it has to be borne in mind that the conditions under which the test was conducted were relatively severe.

Molecular Weight Characteristics of the 95% Ethanol Extract

In the case of this simulant a sufficient mass of extract was obtained with each rubber to record molecular weight data by gel permeation chromatography (GPC).

The GPC results are summarised in **Table 28**. The data shows that this simulant extracts an oligomeric fraction which has a relatively high molecular weight. In both cases the averages are well above the 1000 limit for absorption into the GI tract.

Table 28. Characterisation data obtained on the 95% ethanol extract by GPC		
Rubber Sample	**Mw**	**Mn**
Peroxide cured gum	2,240	1,490
·Platinum cured liquid rubber	1,950	1,410
Mw = weight average molecular weight *Mn = number average molecular weight*		

b) Specific Migration Results

(i) Platinum

The distilled water and 95% ethanol simulants that were in contact with the platinum cured liquid rubber were analysed for their platinum content and the result obtained was less than the detection limit of 0.2 µg/ml, which corresponds to <0.02 mg/dm^2.

(ii) Elements – other than platinum for the liquid silicone rubber

The data obtained using distilled water and 95% ethanol is shown in **Table 29**.

(iii) Formaldehyde

Formaldehyde was determined in the distilled water simulant that had contacted the rubbers for 10 days at 40 °C by EN13130 Part 24-2000. The results were as follows:

Peroxide cured gum rubber   <0.5 mg/kg

Platinum cured liquid rubber   ~0.5 mg/kg

Table 29. Elemental data obtained on the food simulants for the two silicone rubbers		
Element	**Concentration (mg/dm^2)**	
	Distilled water	**95% ethanol**
Aluminium	<0.01	<0.1
Calcium	0.05	<0.5
Magnesium	<0.01	<0.5
Silicon	5.52	4.3

Although not applicable, it can be seen that these values are much less than the 15 mg/kg limit given in the Plastics Regulations (2002/72/EC).

(iv) Silicon

The contact conditions used to prepare the samples were as follows:

Food Simulant	Conditions
Distilled Water	4 at reflux temperatures
95% Ethanol	4 at 60 °C

The results obtained are shown in **Table 30**.

(v) Silicone Oligomers

Table 30. Amount of silicon detected in the food simulant samples		
Simulant	Concentration (mg/dm^2)	
	Peroxide cured gum rubber	Platinum cure liquid rubber
Distilled water	<1.0	<1.0
95% ethanol	6.0	4.3

Aqueous simulant (distilled water) data

The types of silicone oligomer migrants and their approximate levels identified in the distilled water simulant by GC-MS (via an hexane extraction step) are shown in **Table 31**.

Fatty food simulants

The high affinity that silicone rubbers have for fatty food simulants means that the data obtained would have been very similar to that given in the fingerprint section and so no further GC-MS work was carried out.

3. Migration Data Obtained Using Food Products

a) Overall Migration Results

The rubbers were contacted with carbonated water (4 hours reflux) and the results obtained are shown in **Table 32**.

Table 31. Silicone oligomers identified in the food simulant samples		
Siloxane	Peroxide cured gum rubber	Platinum cured liquid rubber
Cyclotetrasiloxane, octamethyl-	39	nd
Cyclopentasiloxane, decamethyl-	52	17
Cyclohexasiloxane, dodecamethyl-	43	23
Cycloheptasiloxane, tetradecamethyl-	15	15
Cyclooctasiloxane, hexadecamethyl-	nd	nd
Total siloxanes extracted from distilled water into hexane (mg/dm^2)	0.15	0. 06
nd = not detected *NB. The hexane extraction efficiency was found to be at or near to 100%*		

Table 32. Overall migration results for the carbonated water food product	
Rubber	Overall Migration (mg/dm^2)
Peroxide cured gum rubber	0.9
Platinum cure liquid rubber	4.8

All of these values are below the EU Silicone Resolution limit of 10 mg/dm^3.

b) Specific Migration Results

(i) Silicon

The rubber samples were contacted with the following food products:

Carbonated water   4 hours reflux
White wine   4 hours reflux
Olive oil   2 hours at 175 °C

The silicon content of the products was then determined and the results obtained were as follows:

(ii) Silicone Oligomers

The carbonated water, white wine and olive oil were analysed to determine if any silicone oligomers could be detected (**Table 33**).

Table 33. Amount of silicon determined in the food product samples		
Food Product	Concentration (mg/dm^2)	
	Peroxide cured gum rubber	**Platinum cured liquid rubber**
Carbonated water	1.7	4.1
White wine	<1.0	0.3
Olive oil	14.5	8.4

The results obtained were as follows:

• Oligomers detected in the carbonated water

The GC-MS data obtained on the two rubber samples using carbonated water is shown in **Table 34**.

• Oligomers detected in white wine

Neither rubber showed an increase in the level of siloxanes compared to the control.

• Oligomers detected in olive oil

The silicone oligomers detected in the olive oil are shown in **Table 35**.

Table 34. Silicone oligomers detected in carbonated water		
Siloxane	**Peroxide cured gum rubber (µg/dm^2)**	**Platinum cured liquid rubber (µg/dm^2)**
Cyclotetrasiloxane, octamethyl-	17.4*	9.4
Cyclopentasiloxane, decamethyl-	14.0*	9.3
Cyclohexasiloxane, dodecamethyl-	8.7	8.0
Cycloheptasiloxane, tetradecamethyl-	6.9	7.0
Total siloxanes in carbonated water (mg/dm^2)	0.05	0.06

Table 35. Silicone oligomers detected in olive oil		
Siloxane	Peroxide cured gum rubber ($\mu g/dm^2$)	Platinum cured liquid rubber ($\mu g/dm^2$)
Cyclotetrasiloxane, octamethyl-	1.1	nd
Cyclopentasiloxane, decamethyl-	13.5	2.8
Cyclohexasiloxane, dodecamethyl-	27.4	8.9
Cycloheptasiloxane, tetradecamethyl-	8.6	2.9
Cyclooctasiloxane, hexadecamethyl-	3.6	0.9
Higher cyclic oligomer	2.2	0.3
Total siloxanes in olive oil (mg/dm^2)	0.43*	0.17*
nd = not detected in the acetonitrile extract		
*Total amounts taking into account the 33% extraction efficiency of the acetonitrile		

4.5.4.4 Overall Summary of the Project Findings

The food migration experiments carried out on cured (both initial and postcure) test sheets of the four silicone rubbers showed that the principal species of interest were siloxane oligomers. The range of species (cyclic oligomers from the trimer to the octamer and linear oligomers in the range n=7 to >20) were detected for all four rubbers.

Overall the levels of siloxane oligomers and other species (e.g., platinum and formaldehyde) that were found to migrate from the rubbers into food simulants was shown to be very low or below the detection limit of the experimental method employed.

Examples of the data obtained on the rubbers are:

a) Platinum <0.02 mg/dm^2

b) Formaldehyde <0.5 mg/kg

c) Total amount of siloxane oligomers – varied from 0.1 mg/dm^2 (distilled water simulant) to 58 mg/dm^2 (95% ethanol simulant)

The values for the siloxane oligomers can be compared to the CoE silicone resolution limit of 10 mg/dm^2. However, testing using 95% ethanol is relatively severe as silicone rubbers are known to have a high affinity for hydrophobic (i.e., fatty) media. Also, analysis by GPC of the oligomers that had migrated into 95% ethanol showed them to have an average molecular weight of ~1,500 Daltons; a significant proportion therefore being above 1000 and so not regarded as being significant in terms of absorption in the GI tract.

Extraction studies on the silicone rubbers using the food products showed that the total amount of silicone oligomers varied from 0.05 mg/dm^2 (carbonated water) to 0.56 mg/dm^2 (olive oil), with the concentration of the most abundant individual oligomer species (e.g., the cyclic hexamer) varying from 0.009 mg/dm^2 (carbonated water) to 0.03 mg/dm^2 (olive oil).

4.6 Published Migration Data

A lot of migration data has been put into the public domain via the research projects funded by the FSA (see Section 4.5). However, a search of the literature reveals that, despite the high profile of food contact materials in general, relatively little has been published in scientific or trade journals concerning rubber products. Some of the principal product and migrant categories that have received attention are covered next.

4.6.1 Food Contact Products

4.6.1.1 Teats and Soothers

A variety of rubbers have been used in the manufacture of teats for baby feeding bottles including natural rubber, silicone and styrene-butadiene block copolymers. Legislation and test methods have been reviewed. Nitrosamines have been a particular cause for concern [19] and over the years efforts have been made to reduce their levels [20, 21]. Legislation on nitrosamines was introduced in 1993 in the form of a European Directive (see Section 3.1).

A study carried out in Japan [22] on the migration of species from teats and soothers made from silicone, natural rubber, synthetic polyisoprene and SBR produced the nitrosamine data shown in **Table 36**.

Predictably, given its origin, the natural rubber based products tended to give the highest results with respect to both *N*-nitrosamines and overall migration and explains why this material is losing ground in this product area to the other materials, particularly silicone.

A survey of the extractables present in rubber teats was published in 1991 [23]. The samples were extracted with diethyl ether or acetone and the extracts analysed by GC

Table 36. Nitrosamines from teat and soother rubbers	
Nitrosamine	Level of migration (ng/g)
N-nitrosodimethylamine	0.6 to 14.2
N-nitrosodiethylamine	1.3 to 11.2
N-nitrosodibutylamine	8.6 to 28.2

and GC-MS. Data was obtained on forty-nine rubber teats commercially available in Holland and a number of compounds not permitted in the Dutch regulations identified, including dibenzylamine, acetophenone, zinc dibenzyldithiocarbamate, 4,4′-thio-bis(2-*tert*-butyl 5-methyl)phenol and bis(2-hydroxy-3-*tert*-butyl 5-ethylphenyl)methane.

A more recent Dutch retail survey [24] looked at the migration of *N*-nitrosamines, *N*-nitrosatable substances and MBT from 19 samples of teats and soothers. In addition to these species, screening work was also carried out for any other potential migrants. The majority of the teats and soothers were found to be based on silicone rubber, and the extractable substances were found to be siloxanes. The rest were natural rubber and only one was found to be above the permissible limits, and that was for nitrosatable substances at 0.23 mg/kg. MBT was only found in one of the natural rubber products and this was below the migration limit of 0.3 mg/teat.

4.6.1.2 Meat Netting

Natural rubber has been the traditional material for elastomeric meat netting for many years and this has led to a number of studies into the levels of *N*-nitrosamines, nitrosatable and other compounds. Work carried out in the USA [25] using a typical product made from natural rubber latex contacted with a 50% ethanol simulant for 150 minutes at 152 °C produced the data shown in **Tables 37 and 38**.

Work in Canada [26] has looked into the levels of certain N-nitrosamines in hams that have contacted natural rubber netting and control samples that have not. The average results obtained on a sample group of twenty products are shown in **Table 39**.

These results illustrate the care that has to be taken in nitrosamine work when devising experiments and interpreting the data obtained.

A collaborative survey of ten meat netting samples obtained from four different manufacturers has been carried out in the Netherlands [27] by workers in the

Table 37. Zinc dithiodicarbonate salts detected in 50% ethanol	
Compound	**Level found (μg/g netting)**
Zinc dibenzyldithiocarbamate	860
Zinc dibutyldithiocarbamate	<26.5
Zinc diethyldithiocarbamate	<26.0

Table 38. Secondary amines detected in 50% ethanol	
Compound	**Level found (ng/g netting)**
Dimethylamine	8.8
Diethylamine	8.7
Dibutylamine	<5.2

Table 39. Nitrosamines detected in netted and control hams						
Sample type	**Nitrosamine compound and level found (ng/g of ham)**					
	NDMA	**NDEA**	**NDBA**	**NPIP**	**NPYR**	**NMORP**
Netted ham	7.3	2.6	<1.4	<4.4	2.1	<1.5
Control ham	4.2	2.0	<1.6	<1.1	1.2	<2.0
NDMA = N-nitrosodimethylamine; NDEA = N-nitrosodiethylamine; NDBA = N-nitrosodibutylamine; NPIP = N-nitrosopiperidine; NPYR = N-nitrosopyrolidine; NMORP = N-nitrosomorpholine						

Inspectorate for Health Protection and Veterinary Public Health and the National Institute of Public Health and the Environment. All ten samples consisted of both natural rubber and vegetable fibres and, in addition to nitrosamines and N-nitrosatable substances, the samples were screened for other potential migrants. Nitrosamines were detected in concentrations up to 2 mg/kg of netting and the two N-nitrosatable compounds dimethylamine and dibenzylamine were found at up to 0.4 mg/kg of netting. These values were not considered to be of concern to public health because of the ratio of meat netting to food product. The other potential migrants identified included alkanes, alkenes, acids, antioxidants, plasticisers and sterols, several of which were not authorised for food contact in the Netherlands, but were allowed in other countries.

4.6.1.3 Rubber Gloves for Handling Food

Datta and Gonlag [28, 29] have recently looked at a number of issues involving latex gloves including nitrosamines, zinc, allergy problems and the food contact legislation and regulations that exist in Europe.

Mutsuga and co-workers have obtained GC-MS data on the extractables that can be obtained from disposable gloves using solvents such as *n*-heptane and *n*-hexane. A paper published in 2001 [30] reported results that were obtained on a range of gloves, including those produced from natural rubber and nitrile rubber. A range of accelerator and plasticiser type species were identified, but it was apparent in the case of the nitrile samples that a relatively large number of extracted compounds could not be identified by GC-MS, no match being found in the commercially available libraries. A second piece of work was then carried out on nitrile gloves [31] to improve the overall quality of the data obtained. Six compounds, which were common to a number of the nitrile gloves used in the original work, were isolated from an *n*-hexane extract by silica gel chromatography and then these compounds were identified by nuclear magnetic resonance (NMR) and high resolution mass spectrometry.

To enhance Malaysia's image as the world's top producer of high quality gloves, the Malaysian Rubber Board banned from January 1st 2005 exports of natural rubber gloves that contain more than 400 µg/g of protein [32]. The ban affected gloves that were designed for food and pharmaceutical use as well as general purpose gloves.

4.6.2 Specific Chemical Migrants from Rubber Compounds

4.6.2.1 Alkylphenol and Bisphenol A

Concerns over their potential to function as endocrine disruptors, led to a Japanese study on the levels of alkylphenols in 60 rubber products [33]. Such compounds are used as starting materials in the manufacture of a number of rubber additives, particularly oligomeric phenolic antioxidants. The work concentrated on four compounds: *p-tert.* butyl phenol (PTBP), *p-tert.*octylphenol (PTOP), *p*-nonylphenol (NP) and bisphenol A (BPA). The results showed the presence of PTOP in three samples in the range 2.2 to 37 µg/g, NP in fifteen samples in the range 2.6 to 513 µg/g, and no PTBP or NP in any samples. Some specific migration experiments for NP were also carried out using water, 20% ethanol and *n*-heptane. The levels were found to vary from 0.004 to 1.519 µg/ml, with the highest results being obtained with the *n*-heptane.

4.6.2.2 Peroxide Breakdown Products

Peroxides are often used to cure silicone rubber and acidic species are among the breakdown products formed. A Japanese study [34] obtained data on such compounds present in extracts obtained from silicone teats and jar seals using thin layer chromatography and UV absorption chromatography. The amount of 2,4-dichlorobenzoic acid in products that had not been postcured varied from 7.7 mg/kg to 24.2 mg/kg; the lowest values obtained using water as the extractant and the highest using *n*-heptane. Postcuring, which is usually carried out for food use silicone products, significantly reduced the levels of this compound.

Peroxides can also be used to cure a number of other rubbers. Work on a peroxide cured nitrile rubber detected between 0.82 and 6.41 mg/litre of the breakdown product diisopropyl benzene in an aqueous food simulant (distilled water) [35].

4.6.2.3 Dimethyl Siloxanes and Other Components from Silicone Rubbers

The food safety aspects of silicone rubbers have been reviewed by Cassidy [36]. In addition to looking at potential food migrants, the review covered a number of areas of potential hazard (e.g., biomedical contact, fire and biodurability) and concluded that silicone elastomers exhibit significantly more benign characteristics than other competing elastomers offered for the same end-use applications.

A test report has been produced by the Fraunhofer Institute of Food Technology and Packaging [37] on the migration of siloxanes from three different silicone rubbers – a high temperature curing material, a room temperature curing material, and a cured liquid silicone rubber. Five different food simulants (iso-octane, ethanol, ethanol/ water, ethyl acetate and olive oil) were used and one of the things investigated was the degree to which the thickness of the sample affects overall migration. In the case of the hydrophobic solvents, it was found to be more important than the polarity of the simulant. As expected, the results obtained with ethanol/water mixtures showed that the amount of migrating oligomeric material reduced markedly with increasing water content, a virtually zero result being obtained above 30%. The migrants were characterised by SFC using both flame ionisation detection (FID) and MS detection. A homologous series of methyl-terminated linear siloxane oligomers up to twenty $SiMe_2O$ units were identified.

A review of the use of a number of analytical techniques (e.g., IR, GC-MS, NMR, atomic absorption (AA) etc.) to identify and quantify polydimethyl siloxanes in a wide range of matrices (e.g., food, pharmaceuticals and cosmetics) has been produced recently. This paper also considers the toxicological issues surrounding polydimethyl siloxanes [38].

4.6.2.4 Accelerators and Antidegradants

An investigation has been carried out [39] on the migration of the accelerator diphenyl guanidine and its reaction products from rubber compounds into food simulants. The concentration of migrants was found to be influenced by fillers present in the rubbers: non-black vulcanisates giving the highest values.

A study carried out in 1999 [40] suggested that when guanidine accelerators are used in sulfur vulcanisation with phenylene diamine based antioxidants, carcinogenic aromatic amines and toxic isothiocyanates may be formed. Similarly, when 6PPD is used as an antiozonant it may decompose and react in aqueous food types to form aromatic amines. The study included the effects that compound formulation and curing conditions have on these reactions.

4.6.3 General Surveys

Two studies have been conducted in France. An early one carried out in 1977, looked at overall migration from a variety of rubbers using a range of test methods [41]. Another survey a few years later [42] was undertaken with a view to preparing a positive list of additives for food contact rubbers.

In 1981 a study was undertaken in Poland on 680 samples of rubber products used in food processing plants [43]. In 35% of the samples migration of metals into 3% acetic acid was reported (14% of the samples contained lead, 2.6% arsenic and 3% barium). The known carcinogen phenyl-beta-naphthylamine was found in 15.1% of the compounds, with amino type antioxidants being detected in 23% of the compounds in total. Poor organoleptic properties were found in 22.1% of the samples. Migration of accelerators occurred in 14.1% of the samples. Overall 97.2% of the rubber compounds did not meet the requirements of the Polish State Institute of Hygiene.

A Polish study has also been carried out in which migration data obtained on Polish produced ether-ester Elitel elastomers was compared with that from a natural rubber and a chloroprene/nitrile rubber blend [44]. When aqueous food simulants that had contacted all of the samples were examined, no phenolic antioxidants, or metals such as arsenic, lead and mercury were found.

A study in Japan [45] looked at the migration of dimethylamine (DMA) into water and hydrochloric acid from 25 rubber articles (including stoppers, chopping boards, spatulas and teats). After 1 hour of refluxing, the water extracts contained 3 to 1280 mg of DMA per kilo of rubber. The study also showed that the thiuram accelerators that were present (TMTD and TMTM) were almost totally decomposed to DMA (a nitrosatable substance). However, in the case of dimethyl dithiocarbamate salts (sodium, zinc, copper and lead examples were included), the decomposition to DMA depended on the solvent used and the salt compound.

4.6.4 Analytical Techniques

Recent papers by Forrest and co-workers and Sidwell have discussed how modern chromatographic techniques such as GCxGC-TOFMS and LC-MS [46, 47] can be applied to the analysis of food contact rubbers. Improvements in the ability to separate, detect, quantify and identify potential and actual migrants are reported.

Barnes and co-workers [48] developed an LC-MS method to identify vulcanisation agents and their breakdown products in food and drink samples. A large sample of 236 retail foodstuffs were analysed for the presence of 2-mercaptobenzothiazole (MBT) and its breakdown product mercaptobenzothiazole (MB). The accelerators 2-mercaptobenzothiazyl (MBTS) and CBS, which are commonly used in food contact rubbers were also looked for. MBT and MB are also known to be breakdown products of these two compounds. The detection limit for these species was found to be dependent on the food product type and ranged from 0.005-0.043 mg/kg. No MBT, MB, MBTS or CBS were detected in any of the samples above these values.

Fichtner and Giese [49] have discussed how both LC-MS and LC-MSxMS can be used to profile the low molecular weight species present in rubber and plastic components that are used in the food and pharmaceutical industries. The advantages that these techniques offer over high-pressure liquid chromatography (HPLC) are emphasised, as is the work that will be required to ensure that these techniques are exploited to their full potential.

Chester and Pinkston have produced a detailed review of the literature on SFC and unified chromatography and how they have been applied to a number of analytical challenges, including food related applications [50].

5 Improving the Safety of Rubber as a Food Contact Material

5.1 Nitrosamines

Nitrosamines form as a result of the reaction of nitrosating agents (e.g., nitrogen oxides in the ambient atmosphere) with secondary amines in the rubber. One of the most prolific sources of secondary amines is a number of the accelerators that are used in sulfur based cure systems; the amines being breakdown products produced as a result of the chemical reactions taking place during vulcanisation. Specific examples of these accelerators, their secondary amine products, and the nitrosamines derived from them are given in **Table 40**.

Table 40. Nitrosamines derived from some commonly used sulfur accelerators		
Accelerator	Secondary amine	Nitrosamine
TMTD, TMTM and ZDMC	Dimethylamine	*N*-nitrosodimethylamine
TETD and ZDEC	Diethylamine	*N*-nitrosodimethylamine
MBS	Morpholine	*N*-nitrosomorpholine
ZDBC	Dibutylamine	*N*-nitrosodibutylamine
DPTD	Piperidine	*N*-nitrosopiperidine

With atmospheric nitrogen being one of the most important nitrosating agents, nitrosamines are easily formed in rubber compounds during both the mixing of the compound and the subsequent fabrication steps (e.g., extrusion and moulding) and they can also be formed during analytical work. A significant amount of work has been carried out by Rapra [51] on the influence of mixing procedures, vulcanisation temperatures, extraction procedures and analysis techniques and the results obtained have shown that a wide variation in nitrosamine levels can be detected in essentially identical compounds.

The main approaches that have been taken to ensure that the amount of nitrosamine (and nitrosatable compounds) in a given compound is as low as possible are as follows:

1) Re-compounding the rubber to wholly or partially substitute accelerators that produce secondary amines with those that do not, e.g., CBS, MBT, MBTS and dithiophosphates [52, 53].

2) Substitution of carbon black filler (or a significant part of it) with white reinforcing fillers (e.g., silica) as carbon black has also been found to act as a nitrosating agent.

3) Switching of the cure system from sulfur based to peroxide based.

4) Use of accelerators that produce higher molecular weight secondary amines, e.g., TBzTD [54].

All of these modifications have to be achieved with retention of both the processing and physical properties of the rubber in mind. The last example, whilst not affecting the levels of nitrosamine formed, reduces their migratory potential.

Because *N*-nitrosamines are suspected of being carcinogenic to humans, all of the major regulatory bodies have laid down limits in teats and soothers. The first to do so were the Germans in 1982, followed by the FDA and Canadian authorities. The EU Directive 93/11 covering the migration of both *N*-nitrosamine and *N*-nitrosatable substances came into force on 1st April 1995. The limits stated in these documents are summarised in **Table 41**.

Although not legally binding, there are also limits for these species that apply to both food and food simulants in the CoE Rubber Resolution. They are:

N-nitrosamines	10 ng/g
N-nitrosatable substances	100 ng/g

Table 41. Regulatory limits for *N*-nitrosamine and *N*-nitrosatable species			
Source	**Extraction media**	**Species**	**Limit (ng/g)**
German	Artificial saliva	*N*-nitrosamines *N*-nitrosatable substances	10 200
FDA	Dichloromethane	*N*-nitrosamines *N*-nitrosatable	10 200
Canadian	Dichloromethane	*N*-nitrosamine *N*-nitrosatable substances	10 200
Dutch	Artificial saliva	*N*-nitrosamines	1
UK	Solvent	*N*-nitrosamines	30
EU	Artificial saliva	*N*-nitrosamines *N*-nitrosatable substances	10 100

The current state of knowledge on the formation of N-nitrosamines in production processes in the rubber industry has been reviewed by Kleps and Pysklo [55]. The scope of the review is wide, taking in the chemistry of their formation, the GC-TEA (GC-thermal energy analyser) method of analysis, the strategies available for reducing their concentration, and a discussion of the relevant legislation.

5.2 Amines

Although not thought to be as potentially harmful as nitrosamines, there are concerns over amines (particularly aromatic amines) and so reducing their concentration is desirable. In common with nitrosamines, they can originate from accelerators, but they also have a number of other important sources, for example from amine type curatives and as the breakdown products of the amine class of antidegradants.

Comprehensive rubber chemistry studies such as those carried out by Rapra over the past 30 years [56] have ensured that the origins of the low molecular weight compounds found in rubbers is well understood and so compounding steps can be taken to avoid those additives that are know to produce amines. Research and development initiatives by additive manufacturing companies have also led to the commercialisation of additives that do not produce amines. An example is a xanthate accelerator produced by Robinson Brothers that breaks down during curing to give only isopropanol and sulfur containing products [57].

5.3 Polyaromatic Hydrocarbons

In the last couple of years PAH, also called polycyclic hydrocarbons (PCH), have been assigned the EU risk phrase R45 (may cause cancer), which means that they are regarded as carcinogenic. This has resulted in a significant amount of attention being given to aromatic process oils and carbon black fillers which are both sources of these species. The manufacturers of hydrocarbon process oils have addressed the issue by altering the production methods so that aromatic hydrocarbon oils could be produced that contained significantly less polyaromatics (i.e., compounds having greater than three benzene rings). These oils have already been used extensively in tyre compounds to reduce PAH deposited in the environment and are also available for food contact products. Food contact regulations, such as those in the USA, have stipulated for many years that for certain types of application (e.g., milk liners) the amount of furnace carbon black could not exceed a given limit (15 pphr) in the rubber compound. Replacement of carbon black with other reinforcing fillers such as silica is another way to ensure that the resulting compound is low in PAH.

5.4 Use of Alternative Compounds

The complexity and versatility of rubber technology has resulted in complicated formulations, often containing up to fifteen ingredients, and this has led to potential problems in terms of the range of species generated which have the potential to migrate into food.

In recent years there has been a trend away from the more traditional rubbers, such as nitrile and EPDM based compounds, to more advanced rubbers having superior properties with respect to heat and chemical resistance, which do not need such a range of additives to ensure that their compounds perform satisfactorily in service. These materials are often simpler in their cure system requirements and so less, and often safer additives (e.g., peroxides), are required.

A rubber which is being used in a number of situations where nitrile and EPDM were once used is butyl rubber (a copolymer of isobutylene and isoprene). Perfectly satisfactory compounds can be made from this material using only five or six ingredients. An example of a novel food-contact product that is being made from butyl rubber is a transportation hose for moving wine between tanks prior to bottling [58].

A further example of a major change in the type of rubber compound used for food contact, is the increasing popularity of thermoplastic rubbers. These rubbers have the advantage of not requiring any cure system ingredients and, in the case of the high performance versions (e.g., polyesters and polyamides) they require less stabilisation than the conventional diene type rubbers that they are replacing. They therefore have the advantage of containing a lower concentration of low molecular weight species in general and none of those that have received the greatest amount of attention due to their potential toxicity. A search of the literature reveals that a significant number of new, food use thermoplastic rubber compounds are coming onto the market.

6 Future Trends in the Use of Rubber with Food

6.1 Increased Use of Thermoplastic Rubbers and High Performance Rubbers

A search of any major abstract database reveals the extent to which new thermoplastic rubbers are entering the food contact market. The attraction of these materials is that they are relatively simple compared to conventional rubbers, which means that they contain less potential migrants overall, and none of those which have received the closest scrutiny in recent years (e.g., nitrosatables and primary aromatic amines). In the main they have entered the market in one of two ways: as replacements for existing rubbers, and as replacements for other food contact materials. There are a number of examples of TPE replacing vulcanised rubbers in the literature. For example, a co-polyester based TPE with high temperature resistance has been developed to replace rubbers such as silicone [59]. A recent development is the manufacture of polyurethane TPE food contact gloves as an alternative to natural rubber or nitrile rubber gloves [60], and there is the continuing utilisation of thermoplastic vulcanisates such as Santoprene as an alternative to conventional vulcanisates [61]. For the manufacture of babies teats and dummies, high performance clear TPEs based on styrene-block copolymers are competing with silicone rubber for this application [62].

An example of TPE replacing other food contact materials is the use of SEBS block copolymers to manufacture synthetic corks for use in wine bottles instead of natural cork [63]. This trend has been assisted in recent years with the continued proliferation of thermoplastic vulcanisates (TPV). These materials differ from traditional thermoplastic rubbers in that the rubbery phase (e.g., the EPDM within a polypropylene matrix) is crosslinked to a degree. The crosslinking of this phase improves a number of the physical and chemical properties of these materials (e.g., temperature resistance and tensile strength) and enables them to compete more effectively with conventional rubbers. The crosslinking is usually carried out during the manufacture of the polymer via a free radical mechanism and so avoids the problem of residual, potentially toxic reaction products.

A novel biocidal thermoplastic elastomer has been produced by modifying a commercial SEBS material [64]. A three-step chemical modification process was carried out involving, firstly a Friedel-Crafts acrylation of the styrene blocks, followed by a hydantoin ring formation, with subsequent halogenation using either chlorine or bromine. The biocidal properties of the material, which could have applications in food packaging and transportation tubing, were proven using several species of bacteria.

The use of 'cleaner' high performance rubbers (e.g., fluorocarbons and halobutyls) is also expected to continue to grow at the expense of the diene type rubbers. The pressure to ensure ever higher margins of food safety is one reason and the demand from owners of food processing plants for increased time between failures is another. Although there are initial cost implications, these are outweighed when complete life cycle costs are considered. For example, a perfluoroelastomer seal may cost 1000 times an equivalent made from EPDM, but it will last long enough in service to easily repay this in reduced maintenance and lost production time.

Efforts are also being made to ensure that there are available high purity and ultra-low contamination versions of the more traditional rubbers, such as EPDM, to ensure that these more cost-effective materials maintain their food use approval status and can be used whenever the service conditions allow it [65].

6.2 Developments in Additives

The move to cleaner and simpler rubbers has enabled a number of controversial additives to be avoided completely and this process is being assisted by the development of new compounds, particularly accelerators, that either produce less potentially harmful breakdown products (i.e., no nitrosatable substances) or ones that have much higher molecular weights and so have a reduced capacity to migrate. An example of the former is the Robac 100 accelerator from Robinson Brothers, which is a xanthate compound producing only isopropanol and sulfur containing breakdown products. An example of the latter is the trend to use benzyl or nonyl thiuram accelerators instead of the traditional methyl and ethyl derivatives.

A particular example of replacement technology that is associated with natural rubber involves the substitution of TMTD in field latex as a preservative with newly developed bactericidal chemicals [66]. The resulting latex can then be used to produce food contact products such as gloves and meat netting thread.

One relatively new class of additive that is finding increasing use in food contact rubber compounds is the antimicrobial agent. One company, Milliken, have recently introduced a family of these compounds based on silver ion exchange resins. They can be used in peroxide cured rubbers such as nitrile rubber and EPDM. The advantage is that they control microbial growth on and within the surface of these rubbers when they are used in food processing lines and this reduces the need for cleaning and part replacement [67]. Another company, Sanitised AG, has also been active in this field and has recently received US Environmental Protection Agency registration for its Sanitised PL 21-60 antimicrobial additive, and has also launched a second product referred to as 'Sanitised Silver' [68].

Rubber sponge products have traditionally been produced using chemical blowing agents that breakdown at the vulcanisation temperatures to produce inorganic gases such as carbon dioxide, nitrogen oxides or sulfur oxides. In the case of silicone rubber, it has also been possible to produce sponge products by the use of physical blowing agents such as low molecular weight organic compounds. There are some toxicity and/or environmental concerns associated with a number of these chemical and physical blowing agents, indeed few have food use approval, and this has been addressed by Dow Corning who have produced a new closed cell sponge technology that uses water as the physical blowing agent [69, 70].

The use of carbon black fillers in rubbers is commonplace, but these additives can contain low molecular weight organic compounds (e.g., aromatics) due to their method of production (i.e., the burning of oil). Carbon blacks have an important influence on both the physical (tensile strength, elongation at break, tear strength, processing viscosity) and chemical (curing rate) properties of rubbers and so extensive development work is required in order to introduce commercially acceptable new products. Due to increasing health and safety and environmental pressure this is beginning to happen. New classes of carbon black are appearing such as the Pureblack class from Columbian Chemicals. This is an ultra-high purity carbon nanoparticle material which is said to combine the properties of traditional carbon black and graphite and, in addition to having the potential to compete with furnace and channel type blacks for use in food contact compounds, to provide a number of processing (e.g., less increase in viscosity per percentage of loading) and final product advantages (better electrical and thermal conductivity) [71].

New food use stabiliser products are being introduced into the market place. Ciba Speciality Chemicals has developed a novel, unique *p*-phenylene diamine antidegradant (Irgazone 997) that has a reduced potential for migration/extraction and as a result is non-staining and low discolouring. It can impart antioxidant, antiozonant and antifatigue properties to a rubber whilst maintaining its ability to achieve food-contact approvals [72]. A new series of *p*-phenylene diamine antiozonants based on 6PPD and 3PPD has also been developed by Ciba Speciality Chemicals [73]. These compounds contain either a hydrophobic thioether or a sulfoxide chain which reduces their solubility in water and hence their tendency, compared to unmodified 6PPD, to migrate into aqueous food simulants.

6.3 Surface Coatings and Modifications

The use of barrier coatings to prevent or eliminate the migration of low molecular weight species has been used for a number of years in the paper and board industry. Recent work has looked at how this type of approach could be used with rubbers. One example studied in Russia [74] compared the migration from a rubber surface that had been fluorinated with a control that had not. Fluorination of the surface was designed to interfere with the 'relay' mechanism of migration where low molecular weight species from the bulk replace those that have been lost from the surface. The results obtained showed that the fluorination was effective in reducing migration from the nitrile rubber, but the degree of reduction was dependent upon both temperature (the reduction being more pronounced as the test temperature approached ambient), and the thickness of the rubber test piece.

The use of antimicrobial additives has been mentioned in Section 6.2, but another route has also been investigated [75], the deposition of silver nanoparticles under formaldehyde radiofrequency plasma conditions onto food grade silicone rubber. The bactericidal properties of the silver-coated surfaces were investigated by exposing them to *Listeria monocytogenes*, with no bacteria being detected after exposure times of 12 to 18 hours.

6.4 Developments in Analytical Techniques

The past five years or so have seen the proliferation of LC-MS instruments to the extent that they have now replaced HPLC in the majority of laboratories. These instruments compliment the GC-MS and with it enable the analyst to routinely generate data on both thermally labile and stable compounds present in rubber products and food up to, and beyond, the GI tract absorption limit of 1000 Daltons.

This, together with the research work that is being carried out on rubbers (Section 4.5), will enable more accurate conformity checks to be performed on compounds, as well as continuing to add to the understanding of the migration behaviour of rubber related species.

In addition to this, development work continues to advance analytical instrumentation so that there are improvements in important parameters such as molecular weight range, detection limits, software assisted peak deconvolution, analysis speed, accuracy of library searching and species selectivity. The advent of multi-hyphenated techniques such as GCxGC-TOFMS and LC-MSxMS are examples of this. These instruments, with their greater resolution and selectivity are also improving the situation with the direct analysis of food products; the large range of low molecular weight compounds in these having caused problems in the past and caused workers to rely quite heavily on food simulants.

7 Conclusion

Since the first edition of this review was issued in 2000, a substantial amount of experimental work has been carried out to investigate and assess the influence that food contact rubbers have on the food products that they come into contact with. The FSA in the UK has been responsible for a lot of progress that has been made in this area, due to its willingness to fund large research projects and the results obtained by these are featured in this report.

Since 2000, considerable progress has also been made towards harmonisation of the rubber in contact with food legislation throughout Europe. The CoE draft resolution on rubber in contact with food was formally adopted at the end of 2004 and efforts will now be made to refine the list of ingredients in the resolution (the inventory list in Technical document No. 1) so that, in time, a Directive may be forthcoming. This is expected to be a long process, however, and may take a number of years.

Another area where notable progress has been made is in the development and proliferation of sophisticated analytical instrumentation. Almost without exception laboratories undertaking food contact work now have an LC-MS to complement their GC-MS, and the major ones will already have second generation versions of these – the LC-MSxMS and GCxGC-MS techniques. Work is progressing well to generate validated methods using these instruments and their presence has enabled considerable advances to be made in the accurate analysis of 'difficult' species (e.g., thermally labile, reactive and oligomeric compounds) in samples resulting from migration experiments. Techniques such as GCxGC-MS are also showing promise in the analysis of food products themselves, as the considerable improvement in resolving power assists the analyst in targeting migrants within such complex matrices. These developments are timely, as it is expected that, as the degree of knowledge concerning the species that may migrate from rubbers increases, and a considerable amount has been learnt in the last five years, more work in the future will concentrate on food that is being consumed by the public and less will be concerned with food simulants.

This review has provided the reader with: an overview of the use of rubber products in the food industry; a summary of the regulatory issues surrounding rubber in contact with food; an insight into the types and scope of the tests that are required to obtain approval; an introduction to the major research projects that have been carried out in the UK in recent years and the results obtained; and a comprehensive survey of the available migration data.

7.1 Sources of Further Information and Advice

There are a number of routes that the researcher can take to obtain further information. It is not possible within this format to provide a comprehensive list, but this section does provide a summary of the key areas where knowledge can be found, with a number of examples included in each category.

7.1.1 Professional, Research, Trade and Governmental Organisations

UK Food Standards Agency

USA Food and Drug Agency

Rubber Division of the American Chemical Society

German Bundesinstitut fur Risikobewertung (BfR) – Federal Institute for Risk Assessment

National Institute of Health Sciences, Tokyo 158, Japan

Institute of Mining, Minerals and Materials (IOM^3)

Rapra Technology

Leatherhead Food International

Central Science Laboratory, York

Pira International

Fraunhofer Institute of Food Technology and Packaging

7.1.2 Commercial Abstract Databases

Polymer Library - Rapra Technology

Chemical Abstracts – American Chemical Society

7.1.3 Key Reference Books and Journals

Rubber Technologist's Handbook, Eds., J.R. White and S.K. De, Rapra Technology Ltd, Shrewsbury, UK, 2001.

Handbook of Elastomers, Eds., A.K. Bhowmick and H.L. Stephens, Marcel Dekker, New York, NY, USA, 2001.

J.A. Brydson, *Rubber Chemistry*, Applied Science Publishers, London, UK, 1978.

R.N. Datta, *Rubber Curing Systems*, Rapra Review Report, Rapra Technology Ltd, Shrewsbury, UK, 2002, **12**, 1.

M.J. Forrest, *Rubber Analysis*, Rapra Review Report, Rapra Technology Ltd, Shrewsbury, UK, 2001, **12**, 7.

Rubbers in Contact with Food, Rapra Review Report, Rapra Technology Ltd, Shrewsbury, UK, 2000, **10**, 11.

Silicone Elastomers, Rapra Review Report, Rapra Technology Ltd, Shrewsbury, UK, 2001, 137.

Migration from Food Contact Materials, Ed., L.L. Katan, Blackie Academic and Professional, Glasgow, UK, 1996.

7.1.4 Food Standards Agency Research Projects

Some of the recent rubber specific research projects funded by the FSA are listed below. Copies of the reports issued at the completion of these projects are available from the Food Standards Agency's library (at Aviation House, 125 Kingsway, London WC2B 6NH).

FS2219   Migration studies – food contact materials

FS2248   Further research on chemical migration from food contact rubber and other elastomers

A03038   An investigation of the breakdown products of curatives and antidegradants used to produce food contact elastomers

A03043   Assessment and quantification of latex protein (LP) transfer from LP-containing contact materials into food and drink products

A03046   Chemical migration from silicones used in connection with food contact materials and articles

Appendix 1

Nineteen Standard Rubber Mixes for the Food Standard Agency (FSA) Rubber Breakdown Products Project

In all cases, phr = parts per hundred of rubber

Compound 20D

General description: an amine cured Hydrin rubber

Ingredient	phr
Hydrin 200	100
FEF N550	50
Maglite DE	7
Zinc oxide	3
Diak No. 1	1.5
Antioxidant MBI	2

Notes

Hydrin 200 = Epichlorohydrin-ethylene oxide copolymer (Hydrin) rubber

FEF N550 = Carbon black filler

Maglite DE = Magnesium oxide

Diak No. 1 = Hexamethylene diamine carbamate (curative)

Antioxidant MBI = 2-Mercaptobenzimidazole (antidegradant)

Compound 48V

General description: an amine cured fluorocarbon rubber

Ingredient	phr
Viton A	100
Maglite DE	15
MT N990	30
Diak No. 1	1.25

Notes

Viton A = Vinylidene fluoride-hexafluoropropylene copolymer (fluorocarbon) rubber

Maglite DE = Magnesium oxide

MT N990 = Carbon black filler

Diak No. 1 = Hexamethylene diamine carbamate (curative)

Compound 49V

General description: a peroxide cured fluorocarbon rubber

Ingredient	phr
Viton GBL 200	100
Zinc oxide	3
MT N990	30
TAC	3
Luperco 101XL	4

Notes

Viton GBL 200 = Peroxide crosslinkable fluorocarbon rubber (composition undisclosed by the manufacturer, DuPont)

MT N990 = Carbon black filler

TAC = Triallyl cyanurate (curative)

Luperco 101XL = 2,5-Dimethyl-2,5(di-tert-butyl-peroxy) hexane (curative)

Compound 50B

General description: a sulfur cured butyl rubber – formulation 1

Ingredient	phr
Butyl 268	100
Zinc oxide	3
Stearic acid	1
Vistanex LM-MS	10
ISAF N220	50
Sulfur	1
Rhenogran MPTD 70	1.43
DOTG	0.3
Nocrac AW	1

Notes

Butyl 268 = Isobutylene-isoprene (butyl) rubber

Vistanex LM-MS = Low molecular weight butyl rubber process aid

ISAF N220 = Carbon black filler

Rhenogran MPTD 70 = DMDPTD = Dimethyldiphenyl thiuram disulfide (curative)

DOTG = N,N´-Di-orthotolyl guanidine (curative)

Nocrac AW = 6-Ethoxy-1,2-dihydro-2,2,4-trimethylquinoline (antidegradant)

Compound 51B

General description: a sulfur cured butyl rubber – formulation 2

Ingredient	phr
Butyl 268	100
Zinc oxide	3
Stearic acid	1
Vistanex LM-MS	10
ISAF N220	50

Sulfur	1
Robac AS-100	0.4
DPG	1
Cytec STDP	1

Notes

Butyl 268 = Isobutylene-isoprene copolymer (butyl) rubber

Vistanex LM-MS = Low molecular weight butyl rubber process aid

ISAF N220 = Carbon black filler

Robac AS-100 = Di-isopropyl xanthogen polysulfide (curative)

DPG = Diphenyl guanidine (curative)

Cytec STDP = Distearyl thiopropionate (antidegradant)

Compound 195S

General description: a sulfur cured styrene-butadiene rubber (SBR)

Ingredient	phr
SBR 1500	100
Zinc oxide	5
Stearic acid	1
HAF N330	70
Sulfur	0.3
Vulkacit Moz/LG	3
Vanax 552	0.75
IPPD	1
Agerite stalite S	1

Notes

SBR = Styrene-butadiene copolymer rubber

HAF N330 = Carbon black filler

Vulkacit Moz/LG = MBS = 2-Morpholinothio benzothiazole (curative)

Vanax 552 = Piperidine pentamethylene dithiocarbamate (curative)

IPPD = Isopropyl paraphenylenediamine (antidegradant)

Agerite stalite S = Mixture of octylated diphenylamines (antidegradant)

Compound 201E

General description: a polychloroprene rubber (CR)

Ingredient	phr
Neoprene WRT	100
Zinc oxide	5
Maglite DE	4
FEF N550	40
Accelerator Thio No. 1	1
Vanax PML	1
Sulfur	0.5
Rhenofit PAN	1
Santoflex 13	1

Notes

Neoprene WRT = Polychloroprene rubber

Maglite DE = Magnesium oxide

FEF N550 = Carbon black filler

Accelerator Thio No. 1 = N,N´-Diphenyl thiourea (curative)

Vanax PML = Di-ortho-tolyl guanidine salt of dicatechol borate (curative)

Rhenofit PAN = N-Phenyl-1-naphthylamine (antidegradant)

Santoflex 13 = 6PPD = N-1,3-Dimethylbutyl-N´-phenyl-para-phenylenediamine (antidegradant)

Compound 238C

General description: a peroxide cured ethylene-propylene-diene monomer (EPDM) rubber – formulation 1

Ingredient	phr
Nordel IP 4520	100
Zinc oxide	5
FEF N550	70
Strukpar 2280	3
Dicup R	2.7
TAC	1

Notes

Nordel IP 4520 = EPDM type rubber

FEF N550 = Carbon black filler

Strukpar 2280 = Hydrocarbon oil process aid

Dicup R = Dicumyl peroxide (curative)

TAC = Trially cyanurate (curative)

Compound 239C

General description: a peroxide cured ethylene-propylene-diene monomer (EPDM) rubber – formulation 2

Ingredient	phr
Nordel IP 4520	100
Zinc oxide	5
FEF N550	70
Strukpar 2280	3
Trigonox 17/40	6.75
TAC	1
Antioxidant 2246	1

Notes

Nordel IP 4520 = EPDM type rubber

FEF N550 = Carbon black filler

Strukpar 2280 = Hydrocarbon oil process aid

Trigonox 17/40 = 3,3-Di-tert-butylperoxy-butanecarboxylic-n-butyl ester (curative)

TAC = Trially cyanurate (curative)

Antioxidant 2246 = 2,2´-Methylene bis(4-methyl-6-tert butyl phenol) (antidegradant)

Compound 240C

General description: a sulfur cured (triple 8 system) ethylene-propylene-diene monomer (EPDM) rubber – formulation 1

Ingredient	phr
Nordel IP 4520	100
Zinc oxide	5

Stearic acid	1
FEF N550	70
Strukpar 2280	3
Sulfur	2
MBT	1.5
ZDEC	0.8
Robac P25	0.8
TMTD	0.8
Rhenofit PAN	1
Aminox	1

Notes

Nordel IP 4520 = EPDM type rubber

FEF N550 = Carbon black filler

Strukpar 2280 = Hydrocarbon oil process aid

MBT = 2-Mercaptobenzothiazole (curative)

ZDEC = Zinc diethyldithiocarbamate (curative)

Robac P25 = Dipentamethylene thiuram disulfide (curative)

TMTD = Tetramethyl thiuram disulfide (curative)

Rhenofit PAN = N-Phenyl-1-naphthylamine (antidegradant)

Aminox = Diphenylamine-acetone condensation product (antidegradant)

Compound 241C

General description: a typical sulfur cured ethylene-propylene-diene monomer (EPDM) rubber – formulation 2

Ingredient	phr
Nordel IP 4520	100
Zinc oxide	5
Stearic acid	1
FEF N550	70
Strukpar 2280	3
Sulfur	1.5
Santocure NS	1.5
ZDEC	0.4
MBTS	1

| Naugard P | 1 |
| Antioxidant 2246 | 1 |

Notes

Nordel IP 4520 = EPDM type rubber

FEF N550 = Carbon black filler

Strukpar 2280 = Hydrocarbon oil process aid

Santocure NS = TBBS = N-Tertiary butyl benzothiazyl sulphonamide (curative)

ZDEC = Zinc diethyl dithiocarbamate (curative)

MBTS = Mercaptobenzothiazole disulfide (curative)

Naugard P = Tris(nonylphenyl) phosphite (antidegradant)

Antioxidant 2246 = 2,2´-Methylene-bis-(4-methyl-6-tert butylphenol) (antidegradant)

Compound 242C

General description: a sulfur donor cured ethylene-propylene-diene monomer (EPDM) rubber

Ingredient	phr
Nordel IP 4520	100
Zinc oxide	5
Stearic acid	1
FEF N550	70
Strukpar 2280	3
Sulfasan R	2
TMTD	0.8
Robac P25	0.8
ZDBC	2
Cytec STDP	1
Rhenofit PAN	1

Notes

Nordel IP 4520 = EPDM type rubber

FEF N550 = Carbon black filler

Strukpar 2280 = Hydrocarbon type process oil

Sulfasan R = Dithiodimorpholine (curative)

TMTD = Tetramethyl thiuram disulfide (curative)

Robac P25 = Dipentamethylene thiuram disulfide (curative)

ZDBC = Zinc dibutyl dithiocarbamate (curative)

Cytec STDP = Distearyl thiopropionate (antidegradant)

Rhenofit PAN = N-Phenyl-1-naphthylamine (antidegradant)

Compound 349N

General description: a peroxide cured hydrogenated nitrile rubber (HNBR)

Ingredient	phr
Zetpol 2000L	100
HAF N330	35
Zinc oxide	3
Perkadox 14/40	6
Antioxidant 2246	1

Notes

Zetpol 2000L = Hydrogenated acrylonitrile-butadiene copolymer (nitrile) rubber

HAF N330 = Carbon black filler

Perkadox 14/40 = 1,3-Bis(tert-butyl-peroxy-isopropyl)benzene (curative)

Antioxidant 2246 = 2,2´-Methylene-bis-(4-methyl-6-tert butylphenol) (antidegradant)

Compound 350N

General description: a sulfur cured nitrile rubber (NBR) containing a bonding agent system for use in fabric-rubber composite products (e.g., conveyor belts)

Ingredient	phr
Breon N36C60	100
Zinc oxide	5
Stearic acid	3
HAF N330	15
Silica VN3	15
Cohedur RS	3
Vulkacit H30	1.5
Sulfur MC	1.5
Vulkacit Moz/LG	0.75
DPG	0.2

MBT	0.2
Wingstay 29	1.43
Santoflex 13	1

Notes

Breon N36C60 = Acrylonitrile-butadiene copolymer (nitrile) rubber

HAF N330 = Carbon black filler

Silica VN3 = Silica filler

Cohedur RS = Resorsinol + stearic acid (bonding agent component)

Vulkacit H30 = Hexamethylene tetramine (bonding agent component)

Vulkacit Moz/LG = MBS = 2-Morpholinothio benzothiazole (curative)

Wingstay 29 = Mixture of styrenated diphenylamines (antidegradant)

Santoflex 13 = 6PPD = N-1,3-Dimethylbutyl-N´-phenyl-para-phenylenediamine (antidegradant)

Compound 351N

General description: a sulfur cured nitrile rubber (NBR)

Ingredient	phr
Breon N36C60	100
Zinc oxide	5
Stearic acid	2
HAF N330	15
Translink 77	15
DOP	5
Sulfur MC	1.5
DPG	0.15
MBTS	1.5
Rhenogran MPTD70	0.29
Rhenofit PAN	1
Wingstay 29	1.43

Notes

Breon N36C60 = Acrylonitrile-butadiene copolymer (nitrile) rubber

HAF N330 = Carbon black filler

Translink 77 = Calcined and surface modified clay filler with vinyl functional surface modification

DOP = Dioctyl phthalate

DPG = Diphenyl guanidine (curative)

MBTS = Mercaptobenzothiazole disulfide (curative)

Rhenogran MPTD70 = DMDPTD = Dimethyl diphenyl thiuram disulfide (curative)

Rhenofit PAN = N-Phenyl-1-naphthylamine (antidegradant)

Wingstay 29 = Mixture of styrenated diphenylamines (antidegradant)

Compound 352N

General description: a sulfur cured NBR/SBR blend

Ingredient	phr
Breon N36C60	50
SBR 1500	50
Zinc oxide	5
Stearic acid	2
HAF N330	15
Translink 77	15
DOP	5
Sulfur MC	1.5
DPG	0.15
MBTS	1.5
Rhenogran MPTD70	0.29
Cyanox 1760	1
Agerite stalite S	1

Notes

Breon N36C60 = Acrylonitrile-butadiene copolymer (nitrile) rubber

SBR 1500 = Styrene-butadiene copolymer rubber

HAF N330 = Carbon black filler

Translink 77 = Calcined and surface modified clay filler with vinyl functional surface modification

DOP = Dioctyl phthalate

MBTS = Mercaptobenzothiazole disulfide (curative)

Rhenogran MPTD70 = DMDPTD = Dimethyl diphenyl thiuram disulfide (curative)

Cyanox 1760 = 4,4-Thio-bis(2-tert-butyl-5-methylphenol) (antidegradant)

Agerite stalite S = Mixture of alkylated diphenylamines (antidegradant)

Compound 503A

General description: an 'efficient' sulfur (i.e., low elemental sulfur level) cured natural rubber (NR) – formulation 1

Ingredient	phr
SMR CV 60	100
Zinc oxide	5
Stearic acid	1
HAF N330	70
Sulfur	0.25
Rhenogran CLD 80	1.25
CBS	0.5
Santoflex 13	1
Nocraw AW	1

Notes

SMR CV 60 = Natural rubber

HAF N330 = Carbon black filler

Rhenogran CLD 80 = Caprolactam disulfide (curative)

CBS = N-Cyclohexyl-2-benzothiazole sulphonamide (curative)

Santoflex 13 = 6PPD = N-1,3-Dimethylbutyl-N´-phenyl-para-phenylenediamine (antidegradant)

Nocraw AW = 6-Ethoxy-1,2-dihydro-2,2,4-trimethylquinoline (antidegradant)

Compound 504A

General description: a conventional sulfur cured NR – formulation 2

Ingredient	phr
SMR CV 60	100
Zinc oxide	5
Stearic acid	1
HAF N330	70
Sulfur	2.5
TMTM	0.4
DOTG	0.6
IPPD	1
Cyanox 1760	1

Notes

SMR CV 60 = Natural rubber

HAF N330 = Carbon black filler

TMTM = Tetramethyl thiuram monosulfide (curative)

DOTG = N,N´-Di-ortho-tolyl guanidine (curative)

IPPD = N-Isopropyl-N´-phenyl paraphenylene diamine (antidegradant)

Cyanox 1760 = 4,4-Thio-bis(2-tert-butyl-5-methyl-phenol) (antidegradant)

Compound Vamac 8

General description: An amine cured ethylene-methyl acrylate (acrylic) rubber

Ingredient	phr
Vamac G	100
Stearic acid	1.5
FEF N550	50
Diak No. 1	1.5
DOTG	4
Antioxidant 2246	1

Notes

Vamac G = Ethylene-methyl acrylate (acrylic) rubber

FEF N550 = Carbon black filler

Diak No. 1 = Hexamethylene diamine carbamate (curative)

DOTG = N,N´-Di-ortho-tolyl guanidine (curative)

Antioxidant 2246 = 2,2´-Methylene-bis(4-methyl-6-tert-butylphenol) (antidegradant)

References

1. A.J. Sidwell and M.J. Forrest, Rubbers in Contact with Food, Rapra Review Report No. 119, Rapra Technology Ltd, Shawbury, UK, 2000.

2. N. De Coster and H. Magg, *Kautschuk Gummi Kunststoffe*, 2003, **56**, 7-8, 405.

3. *Official Journal of the EC: L Series*, 1993, **36**, No.L93, 37.

4. A. Tschech and G. Janeke, *Chemical Weekly*, 2005, L, No.43, 201.

5. J. Sidwell in *Proceedings of the Rapra 10th International Plastics Additives and Modifiers Conference*, Amsterdam, 2004, p.5.

6. A.P. Luning, R. Rijk, H. Zoutendijk and N de Kruijf in *Proceedings of the Rapra Polymer Testing '96 Conference*, Shawbury, UK, 1996, Paper No.4.

7. C. Gueris in *Proceedings of Plastic and Polymers in Contact with Foodstuff Conference*, Edinburgh, UK, 2002, Paper No.1, 12.

8. L. Pysklo, *Elastomery*, 2003, 7, 1, 26.

9. L. Pysklo, T. Kleps and K. Cwiek-Ludwicka, *Elastomery*, 2002, 6, 4-5, 39.

10. R. Sinclair in *Proceedings of Plastics and Polymers in Contact with Foodstuff Conference*, Coventry, UK, 2001, Paper No.6, 3.

11. K. Bouma in *Proceedings of Plastics and Polymers in Contact with Foodstuff Conference*, Coventry, UK, 2001, Paper No.13, 12.

12. *BS 6920-Subsection 2.2.3, Suitability of non-metallic products for use in contact with water intended for human consumption with regard to their effect on the quality of the water - Methods of test - Odour and Flavour of water - Method of testing odours and flavours imparted to water by hoses for conveying water for food and drink preparation*, EN 13130 Part - 24, 2000.

13. *BS 6920 - 2.2.2, Suitability of non –metallic products for use in contact with water intended for human consumption with regard to their effect on the quality of the water, Part 2: Methods of test - Section 2.2: Subsection 2.2.2, Method of testing tastes imparted to water by hoses and Composite Pipes and Tubes*, 2000.

14. *BS 6920 - Subsection 2.2.1, Suitability of non –metallic products for use in contact with water intended for human consumption with regard to their effect on the quality of the water - Part 2: Methods of test - Section 2.2: Subsection 2.2.1 General method of test*, 2000.

15. *BS 6920: Subsection 2.2.1, Suitability of non –metallic products for use in contact with water intended for human consumption with regard to their effect on the quality of the water, Part 2, Methods of test, Section 2.3 Appearance of water.*

16. *BS 6920 - Subsection 2.4, Suitability of non –metallic products for use in contact with water intended for human consumption with regard to their*

effect on the quality of the water - Part 2 - Methods of test - Section 2.4 - Growth of aquatic microorganisms test, 2000.

17. BS 6920 - *Subsection 2.5, Suitability of non–metallic products for use in contact with water intended for human consumption with regard to their effect on the quality of the water - Part 2: Methods of test - Section 2.5: The extraction of substances that may be of concern to public*, 2000.

18. *BS 6920: Subsection 2.5: Suitability of non–metallic products for use in contact with water intended for human consumption with regard to their effect on the quality of the water, Part 2, Methods of test, Section 2.6 The extraction of metals.*

19. C. Cardinet and H. Niepel, *Revue Generale des Caoutchoucs et Plastiques*, 1994, **729**, 64.

20. *Chemical and Engineering News*, 1984, **62**, 38, 31.

21. *Chemical Marketing Reporter*, 1984, **225**, 1, 4.

22. K. Mizuishi, M. Takeuchi, H. Yamaobe and Y. Watanabe, *Annual Report of Tokyo Metropolitan Research Laboratory of Public Health*, 1986, **37**, 145.

23. J.B.H. van Lierop in Proceedings of the Symposium on Food Policy Trends in Europe – Nutrition, Technology, Analysis and Safety, Woodhead, Cambridge, UK, 1991.

24. K. Bouma, F.M. Nab and R.C. Schothorst, *Food Additives and Contaminants*, 2003, **20**, 9, 853.

25. J. Marsden and R. Pesselman, *Food Technology*, 1993, **47**, 3, 131.

26. N.P Sen, P.A. Baddon and S.W. Seaman, *Food Chemistry*, 1993, **47**, 4, 387.

27. K. Bouma and R.C. Schothorst, *Food Additives and Contaminants*, 2003, **20**, 3, 300.

28. R. Datta and A.T. Gonlag, *Gummi Fasern Kunststoffe*, 2003, **56**, 12, 768.

29. A. Gonlag and R. Datta, *Kautschuk Gummi Kunststoffe*, 2004, **57**, 6, 310.

30. C. Wakui, Y. Kawamura and T. Maitani, *Journal of the Food Hygienics Society of Japan*, 2001, **42**, 322.

31. M. Mutsuga, C. Wakui, Y. Kawamura and T. Maitani, *Food Additives and Contaminants*, 2002, **19**, 11, 1097.

32. Malaysian Rubber Board, *Rubber Asia*, 2005, **19**, 2, 115.

33. A. Ozaki and T. Baba, *Food Additives and Contaminants*, 2003, **20**, 1, 92.

34. T. Baba, K. Kusumoto and Y. Mizunoya, *Journal of the Food Hygienic Society of Japan*, 1979, **20**, 5, 332.

35. L.P. Novitiskaya and T.P Ivanova, *Gigiena I Sanitariya*, 1989, 5, 88.

36. L.S. Cassidy, *Progress in Rubber and Plastics Technology*, 1991, 7, 4, 308.

37. O. Piringer and T. Bucherl, *Extraction and Migration Measurements of Silicone Articles and Materials Coming into Contact with Foodstuffs*, FhG Test Report, 1994.

38. K. Mojsiewicz-Pienkowska and J. Lukasiak, *Polimery*, 2003, **48**, 6, 401.

39. *Kauchuk i Rezina (USSR)*, 1978, **1**, 26.

40. H-J. Kretzchmar in *Proceedings of a Rapra Technology Conference on the Hazards in the European Rubber Industry*, Manchester, UK, 1999, 6.

41. *Rapport Technique*, 1978, **106**, 52.

42. *Revue Generale des Caoutchoucs et Plastiques*, 1993, 70, 725, 67.

43. H. Mazur, L. Lewandowska and A. Stelmach, *Roczniki Panstwowego Zaklado Higieny*, 1981, **32**, 2, 97.

44. Z. Roslaniec and H. Ratuszynska, *Polimery Tworzywa Wielkoczasteczkowe*, 1990, 35, 11-12, 450.

45. T. Baba, M. Saito, Y. Fujui, S. Taniguchi and Y. Mizunoya, *Journal of the Food Hygienic Society of Japan*, 1980, **21**, 1, 32.

46. J. Sidwell in *Proceedings of the Rapra 10th International Plastics Additives and Modifiers Conference*, Amsterdam, The Netherlands, 2004, p.5.

47. J. Sidwell in *Proceedings of a Rapra Technology Conference – Rubberchem*, Munich, Germany, 2002, Shawbury, UK, 2002, Paper No.16.

48. K.A. Barnes, L. Castle, A. P. Damant, W.A. Read and D.R. Speck, *Food Additives and Contaminants*, 2003, **20**, 2, 196.

49. S. Fichtner and U. Giese, *Kautschuk Gummi Kunststoffe*, 2004, **57**, 3, 116.

50. T.L. Chester and J.D. Pinkston, *Analytical Chemistry*, 2002, **74**, 12, 2801.

51. K. Scott K and B.G. Willoughby in *Proceedings of Rapra Technology Hazards in the European Rubber Industry Conference*, Manchester, UK, 1999, Paper No.17.

52. M. Saewe in *Proceedings of a Rapra Technology Ltd. Conference – Rubberchem 2004*, Birmingham, UK, 2004, Paper 9.

53. A. Schuch in *Proceedings of Rubberchem '99 Conference*, Antwerp, Belgium, 1999, p.12.

54. R. Datta and T. Mori in *Proceedings of Rubberchem '99 Conference*, Antwerp, Belgium, 1999, p.21.

55. T. Kleps and L. Pysklo, *Elastomer*, 2001, **5**, 5, p.13.

56. M. Forrest and B. Willoughby in *Proceedings of a Rapra Technology Ltd. Conference – Rubberchem 2004*, Birmingham, UK, 2004, Paper No.1.

57. K.B. Chakraborty and R. Couchman in *Proceedings of a Rapra Technology Ltd. Conference – Latex 2002*, Berlin, Germany, Paper No.12.

58. *Revista de Plasticos Modernos*, 2001, **82**, 541, 14.

59. *Plastics Engineering*, 2004, **60**, 11, 6.

60. L. White, *Urethanes Technology*, 2004, **21**, 4, 12.

61. *Plastics Engineering*, 2003, **59**, 9, 18.

62. J. Gu, T. Castile, and K. Venkataswamy in *Proceedings of the Thermoplastic Elastomers Topical Conference: Stretching your opportunities with thermoplastic elastomers*, Akron, OH, USA, 2003, Paper 22.

63. M. Mutsuga, C. Wakui, Y. Kawamura and T. Maitani, *Food Additives and Contaminants*, 2002, **19**, 11, 1097.

64. N. Cliff, M. Kanouni, C. Peters, P.V. Yaneff and K. Adamsons, *JCT Research*, 2005, **2**, 5, 371

65. V.S.Y. Ng and A. Stevens in *Proceedings of the Rapra Technology Conference – Engineering Elastomers 2003*, Geneva, Switzerland, 2003, Session 8, Paper No.26.

66. C. Petri, *Indian Rubber Journal*, 2003, **75**, 28.

67. B. Patel, S. McDowell, R.C. Kerr and G.R. Haas in *Proceedings of the 164th ACS Rubber Division Meeting*, Cleveland, OH, USA, 2003, Paper No.27.

68. *Additives for Polymers*, 2003, 3.

69. R. Romanowski, B.A. Jones and T.J. Netto in *Proceedings of the 164th ACS Rubber Division Meeting*, Cleveland, OH, USA, 2003, Paper No.66.

70. E. Gerlach and F. Giambelli, *Proceedings of the IRC 2003 Conference*, Nuremberg, Germany, 2003, p.339.

71. J. Boyd, *Rubber and Plastics News*, 2004, **33**, 25, 5.

72. J. Wartalski and G. Gnobloch, *Proceedings of the IRC 2003 Conference*, Nuremberg, Germany, 2003, p.389.

73. R.H. Auerbach, C. Boissiere, K. Klein-Hartwig and H.J. Kretzschmar, *Proceedings of the Rapra Technology Conference – Rubberchem 2002*, Munich, Germany, 2002, Paper No.7.

74. A.V. Dedov, V.P. Stoliarov and V.G. Nazarov, *International Polymer Science and Technology*, 2004, **3**, 7, 49.

75. H. Jiang, S. Manolache, A.C.L Wong and F.S. Denes, *Journal of Applied Polymer Science*, 2004, **93**, 3, 1411.

Coatings and Inks for Food Contact Materials

Coatings and Inks for Food Contact Materials

1 Introduction

For many years, Rapra Technology has carried out research projects for the UK Food Standards Agency (FSA). These have covered a wide range of polymer products (e.g., rubbers, silicone-based materials, ion-exchange resins, laminate materials) and have provided the FSA with important information on the materials and manufacturing practices that are used in industry, as well as making an important contribution to the data, via extensive experiments, that is available with respect to the migratory behaviour of these products when they are in contact with food simulants and foodstuffs. This Review Report has, as its origin, an FSA project on *Coatings and Inks* that was carried out at Rapra from 2005 until 2007. The objective of this project was to assess the potential for the migration of substances from coatings and inks that were used in food packaging applications. As a significant amount of work had already been carried out on coatings that were in direct contact with food (e.g., can coatings), a boundary was set that only coatings and inks in non-direct food contact situations would be considered. As the scope of this review report is greater than the Rapra project (see below) and, due to the limitations of this particular format, it has only been possible to include some of the information that was acquired during the course of the FSA project. If the reader has a particular interest in coatings and inks used in these types of applications, they are therefore recommended to apply to the FSA for a full version of the final project report, which was published in March 2007.

Coatings and inks for use with food have been a very topical subject over the last couple of years, mainly due to the culmination of the work that has been carried out by the Council of Europe (CoE). As a result of its efforts, we have seen the adoption of both a *Resolution for Coatings*, and a *Resolution for Inks* used on non-food contact surfaces. The *Inks Resolution* has been controversial with industry bodies throughout Europe, who have claimed that its inventory list is incomplete and not representative of current industry practice (see Section 5.2). In addition to these regulatory developments, this is an active area for research, with a number of innovative and sophisticated products finding commercial applications, e.g., in active and intelligent packaging, and antimicrobials – see (Section 9).

This report has attempted to cover all of the coatings and inks products used in food contact scenarios. Hence, direct and non-direct contact situations are included throughout the food chain, e.g., harvesting, processing, transportation, packaging and cooking. In practice, this encompasses an extremely wide range of polymer systems and formulations, and an emphasis has been placed on coatings and inks used in food packaging, as this is usually regarded as representing the most important application category with respect to the potential for migration to occur. With respect to food packaging, all three of the major material classes are covered, i.e., metal, paper and board, and plastic. In addition to a thorough introduction of the polymers and additives that are used to produce coatings and inks, there are also chapters covering the regulation of these materials, the migration and analytical tests that are performed on them to assess their suitability for food contact applications, the migration data that have been published, and the areas in the field that are receiving the most attention for research and development.

2 Coating and Ink Products for Food Contact Materials

2.1 Polymers for Coatings and Inks

Coatings and inks are polymer-based products, with the polymer being the primary component in the former, and the binder for the pigment system in the latter. In both cases, the two main types of systems are:

1) Those where high molecular weight (MW) polymer is present from the outset – the solvation of the polymer by a solvent, or water, is critical in these cases.

2) Those where the polymer is formed *in situ* from the monomer(s), i.e., the curing types – a number of mechanisms can be responsible for the curing reaction.

Some of the principal types of polymers that are used in coatings and inks products are discussed next. In addition, because there are occasions where conventional rubbers and thermoplastic polymers can be used as protective coatings (e.g., food storage), these are mentioned in Section 2.1.11.

2.1.1 Acrylic

Polyacrylate coatings and binders are based on acrylic or methacrylic esters.

These polymers are created by addition type polymerisations of various combinations of monomers, including:

Methyl esters of acrylic and methacrylic acids

Ethyl and higher esters of acrylic and methacrylic acids

Hydroxyethyl esters of acrylic and methacrylic acids

Diol monoarylates or methacrylates, and

Acid monomers (i.e., acrylic acid and methacrylic acid).

Addition polymerisation is usually achieved by free-radical initiation (e.g., by photochemical processes), and other unsaturated monomers (e.g., styrenics) may sometimes be incorporated for process or product optimisation. In addition, acrylic polymers can be blended with acrylic monomers for viscosity control, which can remove the need for solvents, and polyfunctional acrylic monomers enable crosslinking polymerisation reactions to occur, i.e., curing reactions.

Polyfunctional acrylic monomers can be obtained by reactions of acrylic acid with:

Polyhydric alcohols

Hydroxyl-terminated polyesters

Bis(epoxides) such as bisphenol A diglycidyl ether (BADGE) (see epoxy resins)

The reaction of acrylic acid with polyhydroxy alcohols is a prolific source of polyfunctional acrylic monomers. Examples of di-, tri- and tetra-functional products are:

dipropylene glycol diacrylate (DPGDA)

$$CH_2=CHCO_2(CHMeCH_2O)_2OCOCH=CH_2$$

propoxylated glyceryl triacrylate (GPTA)

$$(CH_2=CHCO_2CHMeCH_2OCH_2)_2CHOCH_2\text{-}CHMeOCOCH=CH_2$$

pentaerythritol tetra-acrylate (PETA)

$$(CH_2=CHCO_2CH_2)_4C$$

Other curing mechanisms exploit the polyfunctionality in hydroxyl groups that may be obtained by copolymerisation with hydroxyalkyl acrylates and methacrylates. Examples of cure via reaction of these side groups include:

Etherification with melamine- or benzoguanamine-formaldehyde resins:

polyacrylic–OH + HOCH$_2$N< → polyacrylic–O–CH$_2$N< + H$_2$O

Urethane formation with isocyanates:

polyacrylic–OH + O=C=N~ → polyacrylic–O–C(=O)–N~

Isocyanate-hydroxyl reactions are particularly active, allowing scope for ambient temperature cures, whilst etherification requires heat and forms the basis of stoving enamels.

Hydroxyl groups are also effective for the ring-opening of epoxides, and therefore epoxy resins can be used for the cure of suitably functional acrylics. The activity of the (amine catalysed) ring opening by hydroxyl groups decreases in the series, ROH > ArOH > RCO$_2$H, and therefore epoxy cures offer useful versatility and control for acrylics containing either hydroxyl or carboxylic acid groups.

2.1.2 Alkyd resins

Alkyd resins are polyesters derived from polyhydric alcohols and mixed acids including dibasic and monobasic types. The polyhydric alcohols include: glycerol (a commonly used compound), pentaerythritol, trimethyolpropane and sorbitol. The dibasic acids (or anhydrides) include phthalic (again commonly used), maleic, isophthalic, adipic and sebacic. The monobasic acids include fatty types with different levels of unsaturation for air-drying performance. These have their origins in natural oils, which themselves usually form the starting material for the production of alkyds. The first stage in production, at an elevated temperature (> 200 °C) is the hydrolysis or alcoholysis (with glycerol), and the second stage is the addition of the dibasic acid component with further heating up to 250 °C.

The most commonly used types of oil include:

a) tung
b) linseed
c) dehydrated castor
d) sunflower
e) soya
f) cottonseed
g) olive
h) coconut

148

Table 1. Types of alkyd		
Type	Oil content (%)	Phthalic anhydride (%)
Short oil resins	35-45	>35
Medium oil resins	46-55	30-35
Long oil resins	56-70	20-30
Very long oil resins	>71	<20

Fatty acids which may be incorporated directly include tall-oil fatty acids and C_8 to C_{10} synthetic types.

A wide range of different products are possible from these reactions. For convenience, alkyds are grouped into four different types, as shown in **Table 1**.

The durability of the cured products decreases with increasing oil content, whereas the long and very long oil alkyds have better brushing properties. Very long oil alkyds form the basis of ink binders. Mineral spirit is commonly used as a solvent for such binder resins, although the presence of alcohols or glycol ethers can also lower viscosity.

2.1.3 Amino Resins (Urea-formaldehyde resins)

Amino resins are obtained from a complex sequence of reactions (e.g., addition, condensation and eliminations reactions) for example:

$$H_2NCONH_2 + CH_2O \rightarrow H_2NCONHCH_2OH$$

$$H_2NCONHCH_2OH + CH_2O \rightarrow HOCH_2NHCONHCH_2OH$$

$$>NCONHCH_2OH + H_2NCONH\sim \rightarrow >NCONHCH_2NHCONH\sim + H_2O$$

$$\sim NHCH_2OH + HOCH_2NH\sim \rightarrow \sim NHCH_2OCH_2NH\sim + H_2O$$

$$\sim NHCH_2OCH_2NH\sim \rightarrow \sim NHCH_2NH\sim + CH_2O$$

A well established procedure reacts urea with a two-fold molar excess of formaldehyde under alkaline conditions to yield an intermediate product mix of mono- di-, and tri-methyolureas, together with some residual urea and formaldehyde. The condensation takes place under acid catalysis, and may be achieved in two stages, the first to create a linear polymer, and the second to achieve a more complete condensation that creates a crosslinked network. Acid compounds such a phthalic anhydride are incorporated into the final curable formulation, the curing being achieved by heating. These, so-called 'unmodified' resins are used for ink binders but are generally regarded as unsuitable for coatings owing to their limited solubility in common solvents.

The solubility limitations can be overcome by modification with alcohols where some of the methylol groups are alkylated, for example:

$$>NHCH_2OH + ROH \rightarrow >NHCH_2OR + H_2O$$

n-Butanol is commonly used for modification in this way, i.e., for 'butylated urea-formaldehyde resins'.

Another variation is the use of melamine (triamino-1,3,5-triazine, $C_3H_6N_6$) instead of urea. The manufacturing sequence is essentially the same as for urea-formaldehyde resins, with addition under alkaline conditions and chain extension and crosslinking under acid conditions. The acid-catalysed etherification is also employed for modification, so that both methylated and butylated melamine-formaldehyde resins find commercial uses. As with the urea-formaldehyde resins, alkylation is used to enhance solubility in common solvents and the alkylated resins are preferred for coating formulations.

Yet another variation is the use of a guanamine instead of melamine – a molecule where one amino group of the melamine is replaced by an alkyl aryl group. One such example is benzoguanamine (2,4-diamino-6-1,3,5-triazine) (**Figure 1**).

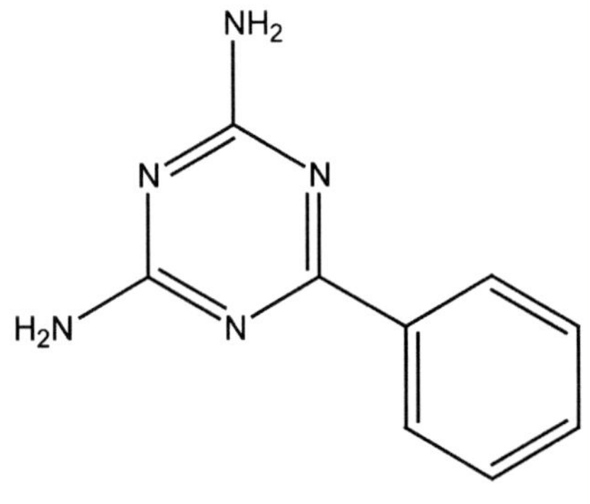

Figure 1. Structure of benzoguanamine

Benzoguanamine closely mirrors melamine in its addition and subsequent condensation reactions with formaldehyde. Benzoguanamine-formaldehyde resins and butylated benzoguanamine-formaldehyde are used in surface coatings.

Any of these amino resins may be used alone, or in combination with other resins, such as epoxies or alkyds.

2.1.4 Epoxy Resins

Epoxy curing reactions exploit the reactivity of the epoxide group. Epoxy resins contain at least two such groups per molecule: BADGE provides the basis of many epoxy resins. BADGE and its homologues are obtained by the base catalysed reaction of epichlorohydrin with bisphenol A (BPA). BADGE (MW 340) is the lowest molecular weight bis-epoxide obtainable from this reaction, i.e., 2,2-bis[4-(glycidyloxy)phenyl]propane.

The phenolic group is also capable of reacting with the epoxy groups present and so this is potentially a polymerising reaction, requiring a significant excess of epichlorohydrin to avoid high molecular weight products. These can be regarded as higher homologues of 2,2-bis[4-(glycidyloxy)phenyl]propane with the inclusion of the $-O-C_6H_4-CMe_2-C_6H_4-O-CH_2-CH(OH)-CH_2-$ repeat unit.

Some polymerisation is helpful, as BADGE itself is a solid, having a melting point of 40-44 °C. The inclusion of a small amount of polymer allows for liquid (at ambient temperature) products.

Epoxies are cured by ring-opening of the oxirane group – either by reaction with active hydrogen compounds or by catalysed homopolymerisation. For coatings, these cures generally fall into two types: ambient temperature cures with polyfunctional amine or thiol co-agents or with polymerisation catalysts; and elevated temperature cures with polyfunctional hydroxyl co-agents (or a combination of epoxide and hydroxyl functionality, such as with anhydrides).

The reactions with active hydrogen compounds are sequential in that the epoxide ring opening generates a hydroxyl group, which, in turn, may initiate further ring opening. This is illustrated in **Figure 2**, for the reaction of a primary amine with epoxy, where both N–H bonds may react with epoxy, as may the hydroxyl groups formed from each addition.

The third reaction shown, i.e., that of the hydroxyl addition, is the form of the reaction where alcohols, or phenols, are used as co-agents.

151

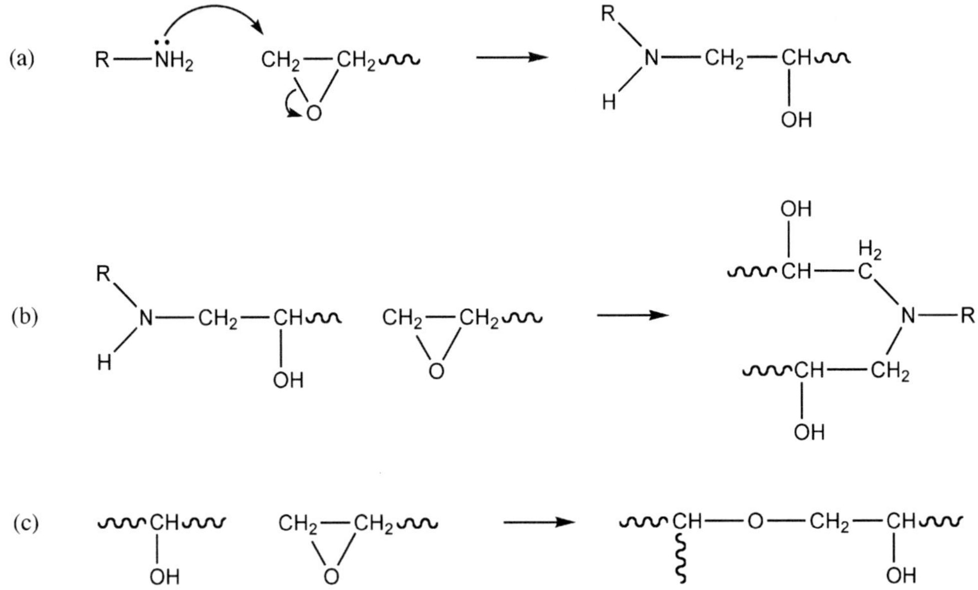

Figure 2. Examples of the reactions that are involved in the curing of epoxy resins

The use of polyfunctional co-agents (i.e., curing agents) is important for network development. Polyalkylene amines such as diethylenetriamine (DTA) and ethylene diamine (EDA), aromatic diamines such as 4,4′ diaminodiphenylmethane (DDM) and aminoamides and their derivatives such as dicyandiamide [$H_2NC(=NCN)NH_2$, called 'DICY'), provide low temperature curing and must usually be applied within a short time (e.g., within 1-2 days) of mixing. A slower reaction affords better control, and hydroxyl-functional curatives form the basis of stoving enamels. Polyfunctionality is obtained with phenol-formaldehyde (PF) resins, where the primary reactions are thought to be those of the phenolic (ArOH) or methylol (ArCH$_2$OH) hydroxyls (depending on the PF resin type used), although other hydroxyl species (including the secondary alcohol formed by ring opening) will also be involved.

Phenol formaldehyde resins offer some of the highest functionalities of epoxy curing agents, and careful selection of resin (PF and epoxy) grades is important if brittle products are to be avoided. In typical phenolic-epoxy stoving enamels, a relatively long-chain bis-epoxide molecule (e.g., higher homologue of BADGE) would be preferred.

The condensation products of this with formaldehyde will contain both amino (>NH) and aminomethylol (>NHCH$_2$OH) groups, depending on the extent of reaction. The inclusion of two, or more, different resins as co-agents in epoxy stoving formulations is not unusual. One benefit may lie in film forming behaviour where the delay of skinning will enhance solvent evaporation to provide for better properties and improved surface gloss.

2.1.5 Cellulosics

Cellulose is a naturally occurring high MW carbohydrate polymer of formula $(C_6H_{10}O_5)_n$. It is extensively hydrogen bonded and possesses remarkable strength/weight characteristics. However, its extensive hydrogen bonding means that it cannot be melted, e.g., for melt processing, without thermal decomposition. Nor can it be dissolved in any solvent in its unmodified form. Given its availability, much effort has been devoted to the development of modified forms.

The hydroxyl functionality in cellulose allows for a number of modifications, notably esterification and etherification. Examples of products produced in this way include:

- cellulose acetate butyrate
- ethyl cellulose
- hydroxyethylcellulose
- methyl cellulose
- nitrocellulose

2.1.5.1 Nitrocellulose

Nitrocellulose is a cellulose ester. It is a highly polar polymer which is easy to dissolve in polar solvents (esters, alcohols) and has good film-forming character on drying. Nitrocellulose (**Figure 3**) is obtained by steeping cellulose (e.g., cotton linters, paper pulp) in a mixture of concentrated nitric and sulfuric acids, at 20-40 °C.

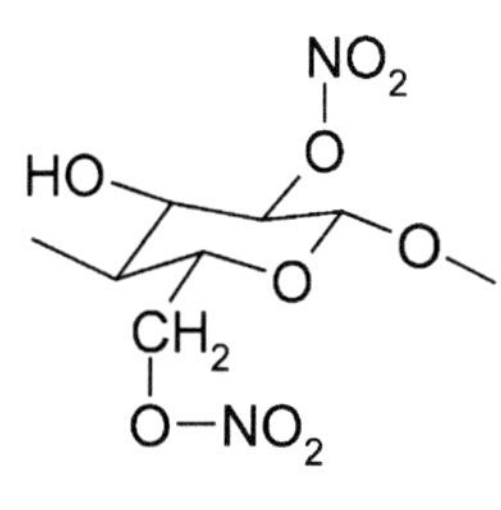

Figure 3. Nitrocellulose (cellulose nitrate) - typical repeat unit

The composition of the acid mix determines the resultant MW. Compositions rich in sulfuric acid provide the lowest MW products, but grades for inks and lacquers (typical molecular weight of 50,000 or lower) are subject to MW reduction by heating with water under pressure at 130–160 °C.

The reactions of production are nitration (esterification) and hydrolysis (of formal –OCH$_2$O–) groups, that is:

$$\sim OH + HNO_3 \rightarrow \sim ONO_2 + H_2O$$

$$>CH-O-CH< + H_2O \rightarrow 2 >CH-OH$$

By reducing H-bonding associations, the esterification reduces crystallinity. Hydrolysis reduces the MW. Typical degrees of esterification for binder polymers are around two nitrate groups per glucose residue, with higher levels being reserved for propellants and explosives.

Dried nitrocellulose has a glass transition temperature (T$_g$) of around 53 °C (depending on the degree of esterification) and therefore requires plasticisation for use in inks and lacquers. Plasticisers for nitrocellulose include the common types (e.g., phthalates, phosphates, etc.) together with natural products such as camphor and castor oil.

Nitrocellulose is relatively water resistant although prone to oxidation. It is not readily amenable to crosslinking, and chemically active drying systems can only be created by blending with other resins. Alkyd, ketone, urea, maleate and acrylic resins are available for formulating with nitrocellulose binders.

2.1.5.2 Other Cellulose Esters

Cellulose acetate is prepared by the acetylation of cellulose (e.g., by first steeping cellulose in acetic acid followed by treatment with acetic anhydride in the presence of sulfuric acid), a process which invariably esterifies all three hydroxyls per repeat unit. The product is cellulose triacetate. The acetate esters have better oxidation resistance than the nitrate esters, but cellulose triacetate is softer and more difficult to plasticise. Acetate levels can be reduced by partial hydrolysis, and the re-introduction of hydroxyl groups can benefit both strength and hardness. Cellulose acetates therefore represent a range of polymers with different degrees of esterification. They are used in films, fibres and lacquers.

Lacquer grade polymers typically have around 2.3-2.4 acetate groups per glucose residue. The degree of substitution is also measured in terms of acetyl content (as a weight percentage) or as the equivalent yield of acetic acid. In the latter description, lacquer grades have 54-56% acetic acid yield (61-62.5% for the triacetate).

Further refinements in synthesis may be achieved by mixed esterification. Cellulose acetate butyrate (CAB) polymers generally offer better mechanical properties than the acetates, and better compatibility with other resins and organic solvents. They can be obtained esterified with a mixture of acetic and butyric anhydrides. Commercial grades for lacquers may have similar numbers of acetate butyrate groups with a very small level of hydroxyl groups (< 0.5) per glucose residue) introduced by a second stage hydrolysis.

2.1.5.3 Cellulose Ethers

Cellulose ethers such as methyl or ethylcellulose are obtained by treating alkali cellulose with the appropriate alkyl chloride (e.g., ROH → RONa → ROMe). The first step is to treat cellulose with 50% aqueous sodium hydroxide at about 60 °C to create the alkali cellulose. The subsequent treatment with the alkyl chloride is accomplished with heat under pressure, the reaction conditions controlling the degree of substitution.

Only a small amount of substitution is needed to disrupt the cellulose structure and generate useful solubility. The solubility characteristics depend on the level of substitution: intermediate levels of substitution by methyl or ethyl groups (e.g., 1.3-2.4 groups per glucose residue) provide solubility in water, whereas higher levels give solubility in less polar solvents (even hydrocarbons at the highest levels of substitution).

Commercial grades of methylcellulose have substitution in the 0.3-1.8 range and are generally exploited for their water solubility. However, higher substitution levels encountered in commercial grades of ethylcellulose where solubility in other solvents is exploited. Ethylcellulose at around 2.5 ethyl groups per glucose residue is compatible with a range of plasticisers and other resins.

For any polymer, solubility depends on molecular weight and, for cellulose polymers, all these treatments are accompanied by a useful degree of molecular weight reduction (of the original cellulose). The treatment which usually delivers the highest levels of water solubility is hydroxyethylation.

2.1.6 Polyesters – Saturated and Unsaturated

Polyesters for hard coatings are commonly based on aromatic diacids e.g., phthalic, isophthalic or terephthalic. MW control (for ease of flow, etc.) can be achieved by reaction with diols in excess. Common examples include low MW glycols such as ethylene glycol, propylene glycol, diethylene glycol, and 1,4-butanediol. Additional monomers can also be incorporated to provide the functionality that is necessary for curing. What these are depends on the cure mechanism used.

Water is a by-product of polyesterification and the extent of reaction is controlled by its removal. This can be hindered by high viscosity mixes, although modifications in reactor design can help. The chemistry of the reaction can also be altered to produce less water, or a different by-product. An example of the latter is the use of the methyl ester of the diacid, and the former by using the anhydride instead of the diacid. With phthalic anhydride, only one mole of water is produced for two moles of ester formed. For example:

~COOCO~ + ROH → ~COOR + HOCO~

ROH + HOCO~ → ROCO~ + H$_2$O

2.1.6.1 Saturated Polyesters

Having a stoichiometric excess of alcohol in the original polyesterification results in a polyester with hydroxyl terminals. The cure of saturated polyesters utilises the reactivity of these end groups. The subsequent cure can be achieved by reactions such as:

Etherification with melamine-formaldehyde, or benzoguanamine-formaldehyde resins:

polyester–OH + HOCH$_2$N< → polyester–O–CH$_2$N< + H$_2$O

Urethane formation with isocyanates:

polyester–OH + O=C=N~ → polyester–O–C(=O)–HN~

Melamine-formaldehyde, or benzoguanamine-formaldehyde resins, are usually polyfunctional in methylol groups, and therefore crosslinking occurs even with difunctional polyesters. The thermal activation necessary for etherification allows for useful control of this cure which is commonly exploited in stoving enamels.

Crosslinking can also be provided via polyfunctionality in the polyester, as is possible when a triol (or polyol), is incorporated into the original polyesterification mix. Examples include glycerol and trimethylolpropane. This allows for crosslinking even with difunctional isocyanates. The hydroxyl/isocyanate reaction is sufficiently active to preclude storage-stable one-component mixes, although these are possible if a blocked isocyanate is used. A blocked isocyanate is a thermally-labile urethane, so that un-blocking is achieved by suitable heating. Phenol is a common blocking reagent, and it is released in the thermal unblocking, for example:

$$RNCO + HOC_6H_5 \rightarrow RNHCOOC_6H_5 \rightarrow RNCO + HOC_6H_5$$

One mole of phenol is released for each equivalent of isocyanate. In effect, the alcohol on the polyester competes with the phenol for the available isocyanate, and the (monofunctional) phenol has the potential to interfere with network formation. However, in thin coatings at elevated temperature, the phenol can be lost by volatilisation: hence phenol blocking is usually reserved for stoving enamels.

2.1.6.2 Unsaturated Polyesters

Unsaturated polyesters have C=C bonds in the backbone, most commonly introduced via maleic anhydride or fumaric acid in the original polyesterification mix. A typical unsaturated polyester can be obtained by heating maleic and phthalic anhydrides with a stoichometric excess (e.g., 20%) of propylene glycol. The mixture is heated at 150-200 °C for up to 16 hours whilst water is continually distilled off. A catalyst, e.g., *p*-toluene sulfonic acid, is sometimes used, and some xylene may be incorporated to assist removal of water by azeotropic distillation.

The cure of an unsaturated polyester is by a free-radical polymerisation, and suitable mobility and reactivity is introduced into the resin by the incorporation of a co-monomer. Styrene is the common choice, although acrylates are also used, and the final stage of the resin formation is the incorporation of this monomer together with an inhibitor such as hydroquinone.

The free radical cure of the resin is essentially a copolymerisation of the unsaturated polymer backbone with the unsaturated monomer. It is possible for oxygen to interfere with the reaction in the curing of thin films and steps to exclude it must be taken. Paraffin wax, which has relatively low solubility in the resin, and hence migrates to the surface, is usually used to create a physical barrier. Its addition, at around the 0.1% level, is accomplished using a small amount of solvent (e.g., toluene) as a carrier. Initiator systems may also be added in solution, these solvents for blending are usually the only solvents used in unsaturated polyester coatings, in which the free monomer provides the major viscosity reduction for application.

Unsaturated polyesters can be cured at room temperature by the use of either two-part initiation systems, or by photoinitiation. The two-part initiation can use either of two approaches to promote peroxide breakdown. One utilises drier chemistry to cause the catalytic breakdown of hydroperoxides. The catalyst used is a metal soap, typically of cobalt, but its action in this case is on intentionally added hydroperoxides, as distinct from the hydroperoxides, formed as a product of air oxidation, that are used in traditional drier chemistry. The hydroperoxides that are added include methyl ethyl ketone peroxide or cyclohexanone peroxide, their breakdown products include methyl ether ketone and cyclohexanone, respectively.

The other approach uses tertiary aromatic amines with acyl peroxides. This involves an electron-transfer process to promote diacyl peroxide breakdown, e.g., with *N,N*-dimethylaniline:

$$ArCOOOOCOAr + C_6H_5NMe_2 \rightarrow ArCOO^- [C_6H_5NMe_2]^+ + ArCOO\cdot$$

but the generation of acyloxy radicals has the potential to produce unwelcome aromatics (e.g., benzene from dibenzoyl peroxide) which may preclude this approach from resins where food contact application is anticipated.

For reference, the anticipated breakdown products of commercial diacyl peroxides are listed in **Table 2** [2].

Table 2. Breakdown products of diacyl peroxides		
Peroxide	**Breakdown products**	
	Hydrogen abstraction	**Radical coupling**
Dibenzoyl	Benzene	
4-Chlorodibenzoyl	Chlorobenzene	Biphenyl 4,4-dichlorobiphenyl
2,4-Dichlorodibenzoyl	*m*-dichlorobenzene	2,2',4,4'-Tetrachlorobiphenyl

2.1.7 Polyurethanes

Polyurethanes (PU) are formed as a result of addition reactions of isocyanates, usually with active hydrogen compounds. The chemistry can be illustrated with respect to hydroxyl addition. The initially-formed adduct itself contains active hydrogens, so that a sequential addition is possible:

$$XNCO + HOR \rightarrow XNHCOOR \qquad \text{urethane formation}$$

$$XNCO + XNHCOOR \rightarrow XNHCON(X)COOR \qquad \text{allophanate formation}$$

This second addition binds two molecules of isocyanate per starting hydroxyl group, so that this sequence makes for a crosslinking reaction. The active hydrogen remains, and the sequence can continue. Hence, depending on the availability of isocyanate, the reaction with allophanate can be represented in a generic sense as an addition polymerisation of the form:

$$nXNCO + XNHCOOR \rightarrow XNHCO[N(X)CO]_nOR$$

Polyurethane curing chemistry is critically dependent on the nature of the co-reagents, their relative proportions, the presence and type of catalysts and the thermal history of the system.

Both aliphatic and aromatic isocyanates may be employed - the former offering better resistance to solar radiation and the latter providing higher reactivities in cure.

For the hydroxyl component of the cure, its activity with respect to isocyanate is governed by both polar and steric effects [3]. Model compound studies show that the reactivity with isocyanate decreases as:

$$CH_3CH_2OH > CH_3OH > (CH_3)_2CHOH > (CH_3)_3OH \sim C_6H_5OH$$

So, in hydroxyl-isocyanate cures, aliphatic hydroxyls provide the co-agents for cure, whilst aromatic (phenolic) hydroxyls provide potential blocking agents. In the latter case the isocyanate is introduced into the cure formulation as is a phenol adduct (phenyl urethane). This adduct dissociates on heating: if the dissociation is in the presence of a more aliphatic hydroxyl then an aliphatic urethane will be formed:

$$X–NHCOOPh \rightarrow X–NCO + HOPh$$

$$X–NCO + HOR \rightarrow X–NHCOOR$$

The aliphatic hydroxyl co-agents in polyurethane cures are usually low MW polymers. Examples include hydroxyl-terminated polyesters and polyethers and hydroxyl-containing acrylics. Even without catalysis, the hydroxyl-isocyanate reaction is sufficiently rapid to compromise storage stability prior to cure and polyurethanes are commonly two-pack systems. Catalysts include tertiary amines, metal soaps and organometallics.

One pack formulations are possible by using blocked-isocyanates (as described above), moisture-curing systems and those incorporating fully reacted, unreacted polyurethanes which cure by alternative chemistry (such as air-drying) or undergo physical drying by solvent evaporation.

Moisture-curing polyurethanes are formulations containing a substantial excess of isocyanate in which the cure proceeds via an initial isocyanate hydrolysis:

$$X–NCO + H_2O \rightarrow X–NH_2 + CO_2$$

This hydrolysis reaction generates active hydrogen, which undergoes sequential reactions (i.e., by urea and biuret formation) in the same manner as for the hydroxyl addition described earlier.

2.1.8 Rosin

Rosin is a natural resin obtained from pine trees. It is a thermoplastic acidic product containing about 90% of so-called 'resin acids' – composed mainly of cyclic isoprenoid acids. The predominant resin acid is abietic acid which has an empirical formula of $C_{20}H_{30}O_2$.

Resins acids are unsaturated and therefore unstable with respect to oxidation. Stability can be improved by hydrogenation or by dehydrogenation (aromatisation) to yield so-called 'modified rosins'. Rosins and modified rosins can be changed further by reactions such as salt formation (e.g., with sodium, potassium or calcium hydroxides) or esterification. Esterification with polyhydroxy alcohols, such as glycerol or pentaerythritol, provides a useful increase in MW.

Rosin esters, rosin salts, modified rosins and modified rosin esters, ranging in physical state from viscous liquids to hard friable solids, are used as binders in a variety of ink formulations. The method of incorporation into the ink depends on the rosin type, notably its acidity. For example, rosin resins may be incorporated as dispersions in linseed oil, as solutions in alcohol or hydrocarbons, or as solutions or dispersions in aqueous ammonia or other alkalies.

Rosin resins are also components of coating formulations, for example as modifiers for alkyds, or epoxies, or for cellulose nitrate lacquers.

2.1.9 Silicone Resins

A silicone polymer has a backbone of alternating silicon and oxygen atoms, i.e., –RR′ Si–O–, where R and R′ may be alkyl, aryl, fluoroalkyl or hydrogen, but are more commonly methyl groups. Therefore, the most common repeating unit in a silicone polymer is dimethysiloxane: $–Me_2Si–O–$.

Silicone resins are crosslinked products. In this case, the crosslinking mirrors other silicon systems (e.g., familiar RTV sealants and elastomers) with the curing process (silanol condensation, i.e., SiOH + HOSi → Si–O–Si) essentially being an extension of the reactions by which the original resin was prepared. Silicone resins are highly branched and are effectively crosslinking systems which are held at an intermediate state of conversion until the reaction can be completed as part of the final application, or fabrication stage. The polysiloxanes for resins are polyfunctional in silanol as a consequence of branch points obtained by the introduction of trifunctional, $RSi(-O-)_3$, or tetrafunctional, $Si(-O-)_4$, units into the siloxane backbone.

In a typical process for silicone resin manufacture, the appropriate mix of chlorosilanes is dissolved in a solvent such as toluene or xylene and then stirred with water. At the end of the reaction, the organic layer is separated and washed free of the acid produced, and then partially distilled to enrich the solids' content of the resulting solution. Whilst further heating or treatment with catalysts may be used to refine the MW distribution, the resin is often kept in this solution until the final cure is required, such as for coating or laminated products.

All resin cures are carried out using heat and a suitable catalyst (metal soaps, organometallics, bases). Heat is essential for progressing a highly crosslinking cure, where the onset of vitrification causes a transition from kinetic to diffusion control. On this basis the temperature of cure should be at least that of the highest temperature expected in service. Although somewhat softer (and less scratch resistant) than other thermosetting resins (epoxies, alkyds), silicone resins are valued for their heat resistance and water repellency.

Heat is also capable of effecting silanol exchange reactions such that the molecular network is, to a degree, in a state of flux. Low molecular weight material can be created through these exchanges, and the principal migratory species in silicone resins are cyclic and linear oligomeric polysiloxanes.

The versatility of silicone chemistry means that it is relatively easy to manufacture silicone resin products that are capable of curing at ambient or relatively low temperatures (e.g., 40 °C), after having been applied to the substrate in a solution form (e.g., in isopropanol) by methods such as dipping or spraying. Such resins are useful in a kitchen and general food preparation environment as they have very good dirt repellent properties and so aid the removal of food residues.

To improve their heat resistance (e.g., up to 650 °C) and chemical resistance, silicone resins are modified by the addition of, for example, epoxy groups. These polymers can be obtained as solutions (e.g., 75% *w/w* in solvents such as methoxypropylacetate) and the coatings resulting from such products, some of which have FDA approval, can be employed as heat resistant decorative coatings in the food industry. It is also possible to incorporate polytetrafluroethylene (PTFE) into such coatings to improve their non-stick performance.

The identity and migration behaviour of species that originate from a variety of food-contact silicone products (rubbers, resins and fluids) are described in the final report of a recent FSA project on silicones managed by Rapra [4].

In addition to being used as the primary binder material, low MW silicone polymers can be modified with organic groups, e.g., phenyl groups, polyether or polyester groups, and used as additives in UV curable coating formulations (e.g., polyester acrylates). Such additives have been shown to enhance the performance of the coatings in a number of ways, e.g., increased temperature stability, superior adhesion to substrates, increased stability towards hydrolysis and a reduction in surface tension [5].

2.1.10 Vinyl Polymers

Examples of vinyl polymers that may be found as binders in inks and coatings, include polyvinyl acetate (PVAc), polyvinyl alcohol (PVOH) and polyvinyl acetals.

Much PVAc is obtained by emulsion polymerisation, a process which delivers the product in latex form. When dried it is brittle, but it is amenable to plasticisation and is soluble in a range of solvents. Although not amenable to subsequent cure, it has potential uses as a binder in aqueous and non-aqueous systems. It is also amenable to internal plasticisation by copolymerisation with a monomer such as 2-ethyhexyl acrylate.

PVAc is the precursor to PVOH. The reaction is usually performed as an alcoholysis in methanol solution (often from a PVAc prepared in methanol solution) in the presence of sodium methoxide. The reaction can be represented schematically as:

$$ROCOCH_3 + MeOH \rightarrow ROH + MeOCOCH_3$$

The degree of substitution can be controlled by the reaction conditions with the most common commercial grades offering around 90% substitution. This level of substitution offers the highest solubility in water, as further hydrolysis increases inter-polymer H-bonding. Fully hydrolysed PVOH is only soluble in water upon heating.

PVOH offers a particularly useful combination of properties: for example it offers good water solubility and a higher level of mechanical performance than normally associated with water-soluble polymers. PVOH is widely used as a binder in water-based inks and coatings.

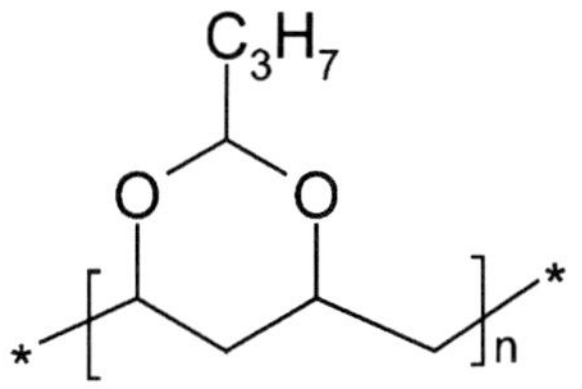

Figure 4. Structure of polyvinyl butyral

Polyvinyl acetals are obtained by condensation of the backbone hydroxyl groups with aldehydes. With formaldehyde, polyvinyl formal is obtained, and with butyraldehyde, polyvinyl butyral is obtained. A typical repeating unit in polyvinyl butyral has the form shown in **Figure 4**.

The precursor to polyvinyl butyral is usually the fully hydrolysed PVOH. This is suspended in ethanol and heated in the presence of butyraldehyde and acid (e.g., H_2SO_4) catalyst, and the product polymer is then precipitated with water. Polyvinyl butyral contains residual hydroxyl groups, and the combination of functionality makes for useful solubility in a range of polar solvents from ketones and esters to glycol ethers and alcohols. To a degree, the properties can be tuned by control of the level of acetylation, but polyvinyl butyrals are generally soft, low melting point polymers.

2.1.11 Other Polymers (Hydrocarbons)

Where relatively thin films of rubber (e.g., ethylene-propylene-diene terpolymer) or thermoplastic polymer (e.g., polyolefin) are used to protect metal surfaces against corrosion (e.g., in storage tanks), they can also be regarded as coatings. In these cases, grades of rubber, or thermoplastics, that are approved for food contact use will be used and the type of polymer (and polymer compound) chosen will depend upon the nature of the food product (i.e., aqueous or fatty), and the physical properties (e.g., abrasion resistance) that the application requires.

In addition to solid polymer products, latex type products are also used as coatings. These are often found on paper and board type packaging for the food industry. For example styrene-butadiene latexes are commonly used for these types of applications, as are terpolymers of acrylonitrile-butadiene-styrene (ABS), the latter offering greater strength and mottle resistance when compared to the former materials.

2.2 Constituents of Coatings

Coatings can be used for protection (e.g., to prevent the rusting of metal) or, in the case of packaging materials, for barrier or release performance. There are also cases where the coating has other functionality, e.g., for anti-mist performance. Coatings that are used in 'release type' applications may be silicone materials, or based on other polymers containing slip additives such as fatty amides. Protective or barrier coatings may be based on curing systems, or evaporating solutions of thermoplastic (i.e., non-crosslinking), or rubber-type, polymers.

The basic ingredients of barrier or protective coatings can be broken down into four main classes: polymers, crosslinking agents, additives and solvents.

The types of polymers that are used to produce coatings have already been discussed in Section 2.1. The remaining constituents are covered next.

2.2.1 Crosslinking Agents

Crosslinking agents are additives, which act chemically on the polymer to effectively create a significant increase in the molecular weight, ultimately producing a complete three-dimensional network. This phenomenon can be regarded as drying via chemical reactions. The choice of crosslinking agent is specific to the polymer concerned and the relevant examples have been discussed with the individual polymer types in Section 2.1.

2.2.2 Other Additives

Additives fall into several types, e.g., pigments, catalysts and various agents to assist flow, leveling or defoaming.

Pigments include the same multitude of types that are discussed for inks in Section 2.3.5, and the catalysts for cure are discussed with the polymer types in Section 2.1. The various other additives include:

i)   Higher-boiling solvents (to those listed in Section 2.2.3) to promote levelling and film formation,

ii) Surfactants to promote surface wetting,

iii) Silicates or metal chelates as thickening agents for viscosity control, and

iv) Esters, soaps waxes or mineral or silicone oils for defoaming action.

164

2.2.3 Solvents

Solvents are not always used. Where they are, they may fall into the following classes:

- Aliphatic hydrocarbons
- Cycloaliphatic hydrocarbons
- Terpene hydrocarbons and terpenoids
- Aromatic hydrocarbons
- Chlorinated hydrocarbons
- Alcohols
- Ketones
- Esters
- Ethers
- Glycol ethers

2.3 Constituents of Inks

Printing inks can be solvent-borne or water borne: they can exploit reactive or non-reactive drying mechanisms. The principal ingredients are 'vehicles' and colorants. The vehicle is the liquid medium, which carries the colorant through the printing process to the substrate and provides stability and gloss to the printed product after drying. Vehicles include solvents, binder resins (i.e., polymers), and additives. Colorants include dyes and, more commonly, pigments.

A range of polymer types, natural or synthetic, may be used as binder resins in inks. Natural resins include rosin types, whilst synthetic binder resins are commonly based on alkyd, amino or acrylic types. Again, the various polymers used are discussed in more detail in Section 2.1.

The inks that are used in food applications usually have complex compositions. The following parameters have to be considered when formulating them:

- Type of substrate
- Type of foodstuffs to be packed
- Type of printing processes and printing equipment
- Package-forming and filling processes

- End user specifications
- Compliance to health, safety and consumer protection regulations
- Compliance with environmental policies for printing, manufacturing processes and end use.

Some of the more important constituents of these inks are described in the next sections.

2.3.1 Solvents

A list of solvents that are typically used in inks is given next:

- Water
- Aliphatic hydrocarbons
- Cycloaliphatic hydrocarbons
- Terpene hydrocarbons and terpenoids
- Aromatic hydrocarbons
- Alcohols
- Ketones
- Esters
- Glycols
- Vegetable oils

2.3.2 Plasticisers

Plasticisers are used in most systems which dry by solvent evaporation, such as gravure and flexography. Plasticisers must be compatible with the resin used, and be present at such a loading to provide flexibility and gloss to the product without impairing application. Nitrocellulose resins are commonly plasticised, and examples of the plasticisers that are used include: phthalates, adipates, phosphates and some vegetable oils.

2.3.3 Driers

Driers are metal soaps which have been traditionally used to catalyse the air oxidation drying of vegetable oils as in alkyd cures. These are typically the naphthenate, octoate or linoleate salts of cobalt, lead and manganese, sometimes with calcium, zinc or barium salts for auxiliary action. The primary drier action is the catalysis of hydroperoxide breakdown, for example:

$$ROOH + Co^{2+} \rightarrow RO\cdot + OH^- + Co^{3+}$$

$$ROOH + Co^{3+} \rightarrow ROO\cdot + H^+ + Co^{2+}$$

Secondary drier action is less clear, but may well involve the formation of ionic crosslinks with any acid groups present in the oil.

2.3.4 Photoinitiators

Photoinitiation is the photoproduction of free radicals or ions capable of initiating a chain reaction such as polymerisation. A photoinitiator is therefore an ingredient which is capable of generating free radicals or ions (usually cations) on the absorption of light (usually UV light).

Cationic photoinitiators are effective for ring-opening polymerisations – as for example with epoxies. Commercial cationic photoinitiators are the so-called 'onium' salts, i.e., diaryliodonium, Ar_2I^+, and triarylsulfonium, Ar_3S^+. Common counter ions include: tetrafluoroborate, BF_4^- and hexafluorophosphate, PF_6^-.

Cationic photoinitiators are a relatively new development. The more traditional free radical initiators are more widely used and generally enable faster curing. Free radical initiators are used with unsaturated monomers, most commonly acrylic. Photoinitiators for radical polymerisation fall into two types: molecules, which are capable of photodissociation and those, termed 'photosensitisers', which generate radicals by an intermolecular reaction. The mechanism of photodissociation can be illustrated with respect to the breakdown of benzoin ethers. Absorption of a UV photon generates an excited state, which can dissipate the excess energy by homolytic bond cleavage:

$$Ph-C(=O^*)-CH(OR)-Ph \rightarrow Ph-C(=O)\cdot + CH(OR)-Ph$$

Carbonyl-containing species which have no easily available route to photodissociation can still function as photoinitiators by the process of photosensitisation. The mechanistic steps for photosensitisation are less well understood than the examples given previously, but are thought to require a co-agent for radical formation. In the case of benzophenone, a hydrogen abstraction reaction by the excited molecule can lead to radical formation:

$$RH + {}^*O=CPh_2 \rightarrow R\cdot + C(OH)Ph_2$$

Some examples of photoinitiators are:

- **Initiators by photodissociation**

  Acylphosphine oxides and bis-acylphosphine oxides

  α-Alkyloxyacetophenones

  α-Aminoalkylphenones

  α-Hydroxyalkylphenones

  Benzoin ethers

  Benzyl ketals

  Halogenated acetophenones

- **Photosensitisers**

  Anthraquinones

  Benzophenones

  Camphorquinones

  Thioxanthones

2.3.5 Colorants

Colorants fall into two main classes: dyes and pigments.

Dyes are coloured substances used for coloration of substrates by processes such as physical adsorption, complex formation or, in some cases, covalent bond formation. Dyes are always soluble in their carriers, and cover a wider colour range than any other colorant class. However, as soluble species (i.e., they operate at the molecular dimension), they are transparent and lack the covering power of pigments.

Dyes have speciality roles in printing. For example, they are used in inkjet printing, in 'lakes' for printing ink pigments and as toners to improve the colour rendering of pigments. Azo dyes form the largest and most versatile class of dyes.

So-called 'direct dyes' (i.e., water-soluble dyes) which attach to substrates by hydrogen-bonding directly from solution are commonly used in inkjet dyes where the substrate is cellulose-based, e.g., paper. Common thermoplastic packaging films (e.g., polyolefins) do not provide a fast surface for these dyes.

Pigments are coloured, colourless or fluorescent particulate solids – usually insoluble in, and unaffected by, the medium ('vehicle') in which they are dispersed. Transparent pigments are used for 'process colours' i.e., colours developed in combination (e.g., cyan + yellow → green). Pigments may be organic or inorganic. The most important class of organic pigments are the azo types. Others include metal complexes (e.g., copper phthalycyanine) and higher polycyclic compounds (e.g., anthroquinone, quinacridone, isoindolinone and perylene).

As pigments are dispersed, and not dissolved, they impart stronger colours than dyes, giving better coverage. Hence, it is pigments rather than dyes that are most commonly encountered as colorants in inks. However, dyes can also be converted into a dispersed form by coating them on translucent particles (e.g., alumina). These are 'lakes' and this process extends the range of colours available as pigments.

2.3.5.1 Azo dyes and pigments

Azo compounds form the most important class of dyes and pigments. They are obtained by 'coupling' reactions between diazonium salts and a range of species such as phenols, arylamines, arylsulfonate salts. This versatility enables a wide spectrum of colours to be manufactured. Members of the naphthalene series couple more readily than single ring compounds, and commercial azo dyes are often multi-ring molecules.

An alternative route to azo compounds is by the oxidative coupling of hydrazones. By these two routes, a huge variety of azo compounds can be produced, including types with one ('monoazo'), two ('disazo') three ('trisazo') or more ('polyazo') azo groups. For example, most of the orange and yellow pigments are disazo types based on substituted benzidines. When each of the coupling components are based on naphthyl or biphenyl derivatives, the resulting disazo, trisazo or polyazo molecules may contain six or more aromatic rings. These large molecules have reduced mobility within the matrix which, in turn, reduces their migration potential.

2.3.5.2 Phthalocyanines

Metal free phthalocyanine $[(C_6H_4C_2N)_4N_4]$, is a macrocyclic compound made up of four isoindole-class units linked by four nitrogen atoms to form a conjugated chain. It is made by heating phthalonitrile (1,2-dicyanobenzene), $[C_6H_4(CN)_2]$, in the presence of bases at 180-200 °C. Phthalocyanine pigments contain metals, such as copper, cobalt, nickel and iron, and provide a wide range of colours through functionalisation of the phthalocyanine ligand itself. Thus, whilst copper phthalocyanine is blue, chlorinated copper phthalocyanine and sulfonated copper phthalocyanine provide two different shades of green.

Copper phthalocyanine may be obtained by heating phthalonitrile and a copper salt in a suitable reaction medium, or by using a phthalic anhydride urea combination in place of phthalonitrile.

Phthalocyanines are stable to atmospheric oxidation at temperatures up to 100 °C, or higher, depending on the metal present, but phthalimide is a possible breakdown product if oxidation occurs.

2.3.6 Other Additives

Other additives that are used in inks include: surfactants, antioxidants, defoamers, biocides, waxes (to prevent setoff and sheet sticking), thickeners (e.g., cornstarch), anti-skinning agents (to prevent premature drying on rollers – could be an antioxidant). Monomers (e.g., acrylic) are additives in reactive systems.

3 Coatings and Inks used in the Food Chain

The food chain is becoming increasingly complex, with an increasing choice for the consumer due to the proliferation of pre-prepared foods, rapid movement of packaged goods by road, and the increasing competitiveness of the global market.

There is a net flow of materials along the food chain from the farmers and growers, through the manufacturers and processors, to the major supermarkets and food service outlets, e.g., caterers.

This section highlights some distinct areas in the food production chain, and the types of coatings and inks that are used in those areas. Packaging has been given the most attention as it is the most important area due to a number of factors. It is the area in which the greatest number and range of coating and ink products are used in, and it also represents the highest potential for migration to occur as the contact times and areas (per unit of food) are the usually the greatest.

3.1 Food Packaging

3.1.1 Packaging Types

It is convenient to divide the packaging that is used for food into three generic categories:

a) **Metals**

Coatings and inks can be used in the following sectors of metal packaging:

- Beverage cans and ends
- Food cans and ends
- Caps and closures
- General line (standard, non-specific manufacturing line) used for dry food products
- Aerosols and collapsible tubes used for food products

b) **Flexibles (i.e., plastic)**

Coatings, adhesives, primers, varnishes and heat seals are used in multi-constructions of plastic films and foils

c) **Paper and Board**

Coatings, adhesives, primers, varnishes and heat seals are all used in this sector.

Some approximate data on the size of the market for these types of products is given in **Tables 3 and 4**.

Table 3. Current size of the UK and European markets for metal packaging
European market (billions of cans)
40
29
30
5.5
7

Table 4. Current Size of the UK non-metal packaging market			
	Inks (Te)	Coatings/sealants/ varnishes (Te)	Adhesives (Te)
Flexible packaging	7200	1500	5000
Paper/board	2000	4000	4000
Food labels	500	1000	1500

It is useful to try and illustrate the relative importance of the different polymer types used in the production of metal, flexible and paper and board packaging. This has been attempted in **Tables 5 to 14** [6]. In these tables, the polymer types have been given a rating from 0 to 10. This denotes the relative importance of the polymers, with the least important being given a rating of 0 and the most important a rating of 10.

Table 5. Polymers used in metal packaging - beverage bodies and easy open ends			
Polymer type	**Relative importance (1-10)**	**End use**	**Comments**
Alkyds	0-1	None to very little	Polyesters preferred
Acrylics	7	Ends and bodies	Very common base resin for w/b coatings and varnishes
Polyesters	10	Ends and bodies	Higher quality base resin than acrylics for coatings/varnishes
Epoxies	9	Ends and bodies	Significant use as base resin for w/b externals of ends
Urethanes	1	None to very little	Food contact issue even for externals
Vinyl polymers	2	End stock	Chlorine containing vinyl resins being withdrawn
Phenolics	2	Ends and bodies	Crosslinking or colouring resin
Amino	4	Ends and bodies	Principal crosslinking resin for epoxy resin for clear/coloured coatings
Acrylates	1	Base of bodies	Small interest for acrylate UV cured rim varnish
Cellulose	<1	None to very little	Maybe only as an additive
Hydrocarbon	0		Maybe only as lubricant
Others	1	Lubricants	Small volumes of lubricants and additives are used including PE, PTFE, Carnauba wax and lanolin
w/b = water based			

Table 6. Polymers used in metal packaging - external food bodies and ends including welded three piece/drawn/DRD and DWI cans and ends			
Polymer type	Relative importance (1-10)	End use	Comments
Alkyds	1	Very little	Being replaced by polyesters
Acrylics	2	Overprint varnishes	Provides good clear glossy finish
Polyesters	3	Ends	Growing use crosslinked with amino, phenolic and isocyanate
Epoxies	10	All sectors	Main base resin for ends(standard and EOE) and DWI washcoat
Urethanes	3	ROPP and other caps	Flexible and tough technology
Vinyl polymers	2	Diminishing use	Were used on DRD cans – replaced by polyesters and epoxies
Phenolics	6	All sectors	Important crosslinker for gold coatings
Amino	4	All sectors	Crosslinker for clear and white coatings/varnishes
Acrylates	1	Three piece bodies	Basis for UV curable varnishes and also flow agents
Cellulose	<1	Small	Maybe as an additive
Hydrocarbon	<1	Small	Maybe as an additive
Cycloaliphatics	1	Three piece bodies	Base resin for cationic UV curable varnishes
Others:			
PTFE/PE	<1	All sectors	Lubricants used to aid post forming
Carnauba	<1	All sectors	Lubricants used to aid post forming
EOE: Easy open end			

Polymer type	Relative importance (1-10)	Vacuum closures (including: RTO/ PT/ lug type/)	Roll on pilfer Proof	Crown corks	Comments
Alkyds	1	Yes		Yes	Combined with vinyl to give good flexibility-being replaced by better polyester technology
Acrylics	1	Yes		Yes	Far greater use in Europe
Polyesters	10	Yes	Yes	Yes	Main resin type preferred for flexibility and good heat resistance and colour retention
Epoxies	4	Yes		Yes	Base of traditional epoxy ester technology being replaced by polyester base limited flexibility
Urethanes	6	Yes	Yes	Yes	Ideal for the more extreme draw of ROPP but used for all
Vinyl polymers	2	Yes		Yes	Combined with alkyd technology, but now being replaced
Phenolics	2	Yes		Yes	Crosslinker for gold epoxy -based clear externals
Amino	6		Yes	Yes	Crosslinker for polyester technology
Acrylates	1	Yes		Possible	Base for UV technology with limited draw. Proven for RTO products
Cellulose	0				
Hydrocarbon	0				
Others:					
Cycloaliphatic	1	Yes	No	Possible	Base for cationic varnishes
PTFE/PE	<1	Yes	Yes	Yes	Lubricants, used in small quantities

Table 7. Polymers used in metal packaging - Caps closures to include vacuum closures/RTO/ROPP/PT/crowns etc.

Table 8. Polymers used in metal packaging - General line (including, packaging for dry foods, e.g., baby food, coffee and confectionery)

Polymer type	Relative importance (1-10)	End use	Comments
Alkyds	1	Size coat	Sizeless systems now preferred
Acrylics	5	Base for white coatings and varnishes	Good heat resistance and appearance
Polyesters	10	Base for white coatings and varnishes	Excellent combination of properties and crosslinked with amino or urethanes
Epoxies	2	Traditional base for epoxy ester varnishes	Epoxy esters prone to yellowing
Urethanes	4	Superior technology	Combination with polyester
Vinyl polymers	1	Historical base for size/white and varnish	Vinylalkyds being replaced by polyesters or acrylics
Phenolics	1	Used as crosslinker in clear golds	Combination with epoxy
Amino	4	Clear transparent or white basecoat	Combination with polyester
Acrylates	5	Basis for UV free radical varnishes	UV technology replacing thermal technology in UK
Cellulose	0		
Hydrocarbon	0		
Others:			
Cycloaliphatic	4	Basis for cationic UV varnishes	More flexible cationic technology is more versatile than free radical. Replacement of conventional set to continue

Table 9. Polymers used in metal packaging - Collapsible tubes, three piece steel and aluminium monobloc aerosols					
Polymer type	Relative importance (1-10)	Collapsible tubes	Three piece aerosols	Monobloc aerosols	Comments
Alkyds	<1		Yes		Mainly replaced by polyesters
Acrylics	4		Yes	Yes	Lower cost option to polyesters
Polyesters	10	Yes	Yes	Yes	Dominates the external size/white and clear basecoats and overprint varnishes
Epoxies	3		Yes	yes	Used more on three piece aerosols than monoblocs
Urethanes	6	Yes	Yes	Yes	Isocyanate capped PU used extensively with polyesters
Vinyl polymers	0-1		Possible		Could be still used on cones and domes
Phenolics	2		Yes		Crosslinker/tinter for epoxy-based gold coatings
Amino	5		Yes		Clear crosslinker alternative to polyurethane
Acrylates	1		Yes		Base for UV curable varnishes
Cellulose	0				
Hydrocarbon	0				
Others:					
Cycloaliphatic	2		Yes	Possible	Base for the more flexible and better adhering UV technology for varnishes
PTFE/PE	<1	Yes	Yes	yes	Small volume used as internal lubricants
Carnauba	<1	Yes	Yes	yes	Small volume used as internal lubricants

Table 10. Polymers used in flexible packaging - adhesives used for laminating Plastic films/foil/paper and metallised

Polymer type	Relative importance (1-10)	Solvent-based	Solvent-free	Water-based	UV or EB cure	Comments
Alkyds	0					
Acrylics	2			Yes		Acrylic functionality aids water dispersability
Polyesters	7	Yes	Yes			Used as backbone for high performance adhesives or curatives
Epoxies	2	Yes		Yes		Used in conjunction with amine curatives
Urethanes	10	Yes	Yes	Yes		Polyether or polyester urethane technology is the most widely used adhesive technology in the UK and Europe
Vinyl polymers	0					
Phenolics	0					
Amino	0					
Acrylates	1				Yes	Could be background resin for UV or EB technology
Cellulose	0					
Hydrocarbon	0					
Others:						
Amine	2	Yes		Yes		Used in conjunction with an epoxy
Polyether/ polyol	6	Yes	Yes	Yes		Can either be used as part of the PE/PU backbone in the adhesive part or as a curative in its own right with an isocyanate capped PU
Silane/silanol	4	Yes	Yes	Yes		Used as an adhesion promoter in the curative
Amide	1	Yes	Yes	Yes	Yes	Erucamide or oleamide-based lubricants used to adjust bond strengths

Polymer type	Relative importance (1-10)	End use	Comments
		Table 11. Polymers used in flexible packaging - Primers/heat seals and varnishes	
Alkyds	2	Varnishes	In conjunction with amino resins for heat resistance
Acrylics	5	Lidding heat seals	Can be applied as waterbased in conjunction with vinyls e.g., yoghurt pots
Polyesters	3	Heat seals	Good performance
Epoxies	2	Sterilisable trays/coloured cartons	Can be used with phenolics and thermoplastic dispersion e.g., PP base for coloured coatings for drawable trays
Urethanes	4	Varnishes	High chemical and heat resistance
Vinyl polymers	6	Lidding heat seals or barrier cartons	Traditional polymer base for heat seals-polyvinylidere chloride major barrier coating
Phenolics	2	Sterilisable trays	Crosslinker for epoxy based coatings and/or heat seals
Amino	2	Varnishes	Used in conjunction with alkyd and polyester resins
Acrylates	2	Varnishes	Low heat resistance
Cellulose	3	Varnishes	Nitro cellulose, low cost and low temperature drying
Hydrocarbon	0		
Others:			
Polyamide	5	Cold seals	Release additive
Ethylene vinyl acetate	5	Heat seals/lidding	Water-based EVA replacing high VOC solvent based systems
Maleinised PP	3	Sterilisable trays	Used as thermoplastic dispersion with thermoset epoxy phenolic
Polyethylene-imine	1	Primer for extrusion lamination	Improves adhesion for direct extrusion lamination
Natural rubber	5	Cold seals	Provides cohesive strength
Stearates/soaps	1	Anti-mist Coatings	Absorbs water and allows clear view of products packed
EVA: Ethylene vinyl acetate			

Table 12. Polymers used in inks for metal packaging			
Polymer type	Relative importance (1-10)	End use	Comments
Rosin esters	0		
Hydrocarbon resins	0-1	Diluent	Only as a minor diluent, e.g., tridecanol or 260-290 °C hydrocarbon distillate
Phenolics	0		
Cellulose resins	0		
Acrylic	0		
Vinyl resins	0		
Polyamide	0		
Polyurethane	0		
Amino	2	Beverage externals	Used to crosslink polyesters
Alkyds	3	All metal packaging sectors	Conventional ink systems not requiring crosslinking
Ketone resins	0		
Polyester	10	1. Beverage external decoration 2. General line/aerosols	1. Crosslinked with amino 2. Used for polyester acrylate in UV
Epoxy	4	All metal packaging sectors	Used for epoxy acrylate in UV.
UV/EB curable polymers	7	All metal packaging sectors	Free radical UV technology based on mainly epoxy acrylate with some polyester acrylate
PVB	0		
Silicone/ silicate	0		
PTFE	1	Lubricant	Provides necessary slip to the surfaces
Titanates/ chelates/ maleic resins	0		

Table 13. Polymers used in inks for paper and board			
Polymer type	Relative importance (1-10)	End use	Comments
Rosin esters	0		
Hydrocarbon resins	1	Diluent	Only as a minor diluent in conventional inks
Phenolics	0		
Cellulose resins	0		
Acrylic	0		
Vinyl resins	0		
Polyamide	0		
Polyurethane	0		
Amino	0		
Alkyds	4	Conventional inks	20% of the food paper/ board market
Ketone resins	0		
Polyester	5		Resin used in conjunction with acrylate for UV
Epoxy	5		Resin used in conjunction with acrylate for UV
UV/EB curable polymers	10	Mainly UV Some EB at Tetrapack	80% of market Combination of polyester and epoxy acrylates
PVB	0		
Silicone/silicate	0		
PTFE/PE	0-1	Lubricant	
Titanates/chelates maleic resins	0		

Table 14. Polymers used in inks for flexible plastic packaging			
Polymer type	Relative importance (1-10)	End use	Comments
Rosin esters	0-2		Minor use in gravure
Hydrocarbon resins	0		Minor use
Phenolics	0		Not of interest due to formaldehyde concern
Cellulose resins	10		Nitro cellulose is used in almost all inks applied via the flexography and gravure processes
Acrylic	2	Water-based only	Mainly in USA
Vinyl resins	6		Move to withdraw chlorine containing polymers
Polyamide	6		Odour issue from amine and solvents possible
Polyurethane	10		High performance ink media reducing the NC content
Amino	0	Mainly coatings	Mainly two pack odour issue
Alkyds	0		
Ketone resins	1	To modify NC	
Polyester	0		
Epoxy	0-1		Used in small quantities as an adhesion promoter
UV/EB curable polymers	2		Acrylated PU can be used for UV curable systems
PVB	5	Preferred base resin system for confectionary/ cold seal	Not suitable for heat sealable end uses
Silicone/silicate	0-1		Waxes
Titanates/chelates/ maleic resins	5	Adhesion promoters	Essential additives for inks
NC: Nitrocellulose			

The preceding Tables can be broken down as follows:

1) **Tables 5 to 9**   Coatings used in Metal Packaging
2) **Tables 10 and 11**   Coatings and Adhesives* used in Flexible Packaging
3) **Tables 12**   Inks for Metal Packaging
4) **Tables 13**   Inks for Paper and Board Packaging
5) **Tables 14**   Inks for Flexible Packaging

*Adhesive layers, such as the 'tie layers' in laminates, can be regarded as a coating and so this type of product has been included here.

The information in these Tables is complemented by the discussion in Sections 3.1.2 to 3.1.7.

3.1.2 Coatings Used in Metal Packaging (Tables 5 to 9)

3.1.2.1 Beverage Body and End Stock Externals (Coatings Used in Conjunction With Inks)

The UK market for alcoholic drinks is dominated by metal packaging whereas the soft drink market mainly uses polyethylene terephthalate (PET) bottles. Glass is only a small market for each (< 10 %). Body stock in the UK consists mainly of aluminium construction with a small proportion made from tinplate and this has an impact on the coatings and inks chosen.

The preferred combination on aluminium can stock is polyester-based inks (see **Table 12**) with a water-based over varnish based on polyester or a polyester/acrylic technology.

In Europe, countries with a strong steel industry tend to produce more beverage bodies from tinplate. The coating situation with tinplate is quite different to aluminium. Generally, a water-based, white basecoat, based on a polyester/acrylic combination, acts as an excellent base for inks requiring no over varnish. Known as 'Novar' inks in the trade they contain higher levels of lubricants to provide necessary slip and abrasion resistance through the filling lines and track work.

Table 5 shows the strong use of water dispersible polymers in varnishes and pigmented coatings, e.g., acrylic, polyester and epoxy.

Can end stock is made of aluminium and the externals are primarily solvent-based epoxy amino technology. A very small percentage of ends will be decorated for promotional reasons. The vast majority of beverage ends use clear, gold or coloured external coatings. The coatings are applied using high speed coil application.

The UK market primarily produces three piece welded food cans (80%) with approximately 20% produced using the drawn and wall ironed (DWI) process.

A major proportion of three piece welded bodies are left uncoated on the exterior; no external decoration being needed as paper labels are used. However, the steel ends will always be coated with an epoxy-based material. Many of the easy open ends will be printed with opening instructions and others may be printed for promotional reasons.

Water based epoxy wash coat dominates the external protection of DWI bodies; again no inks are used as paper labels are preferred.

A small number of drawn steel 'pie cans' are produced in the UK which will receive an external combination of white basecoat, inks and varnishes.

More flexible coatings are used on both aluminium and steel draw redraw (DRD) cans to aid in the forming process. In addition, the Europeans still tend to decorate their cans with basecoats/inks and varnishes. In Europe the emphasis is more on quality and therefore direct printing of cans and ends is preferred rather than using paper labels. **Table 6** shows the use of various other resins for these cans.

The inks used on these products will follow the polymer combinations shown in **Table 12**. Generally, conventional inks are used on food cans and ends which are based on polyester technology.

This market is quite diverse and there are many different types of caps and closures produced. The most prevalent type of closure is the steel vacuum closure which is used on glass jars for baby food, jams and pastes, etc. Various constructions are included such as Regular twist off (RTO) and Press Twist (PT) etc., on processed food glass jars. These external fittings use mainly use polyester or epoxy base coats in combination with inks and varnishes.

Caps that are used on alcoholic glass bottles are of the aluminium Roll on Pilfer Proof (ROPP) construction. They require very flexible basecoats, inks and varnishes to withstand the extreme forming process. Typical choices will be based on polyester-polyurethane technology.

Crown corks are still used on premium beer and beverage glass bottles and again use combinations of polyester basecoats, inks and varnishes based on polyester or epoxy ester technologies.

Only a small proportion of general line cans produced in the UK are used for food products, with the vast proportion of general line cans being used for paints, wood varnishes and other non-food consumer products. The main food packaging use for general line cans is for decorated confectionary boxes, decorated boxes or cylinders for premium bottles of spirits, or for baby food powders.

In the UK, in the last few years, there has been a considerable move away from thermally cured coatings and inks for these products to using both UV curable inks and varnishes. This is inspired by volatile organic compounds (VOC) regulations, quality and economics. The coloured basecoats, however, are normally solvent-based and produced from a range of resins, but mainly polyesters.

For the varnishes, UV technology is either free radical based (acrylate-based) or the higher performance cationic technology based on cycloaliphatic resins.

The vast majority of aerosols produced in the UK are of welded tinplate construction. A small proportion of the total are produced using the Monobloc aluminium process. Both types of construction use direct print onto a base coated body which is subsequently varnished, with the mono bloc aluminium aerosols mainly using conventional solvent borne basecoats, and conventional inks and varnishes based on flexible polyester technology.

The main three piece tinplate producers are now moving to UV curable inks and varnishes. Again, solvent base-polyester or epoxy resins are used for the basecoats.

Aluminium collapsible tubes are included in this table, although as none are manufactured in the UK, the relatively small number used for pharmaceuticals and food pastes are imported mainly from Europe. This type of packaging has been mainly replaced by rigid plastic polyethylene (PE) tubes.

3.1.3 Coatings and Adhesives for Flexible Packaging (Tables 10 and 11)

The types of flexible packaging in the market can be broken down as follows [7]:

a) General purpose

b) Medium performance

c) High performance

Examples of each of these three are described in the next sections.

General Purpose

This type of packaging is typically used for products such as snacks and crisps. In addition to the metalised inner plastic layer (~18 µm thick) and outer plastic layer (again ~18 µm thick), there will be a 2 µm adhesive layer and a 5 µm print layer. The presence of the metallised layer provides a degree of oxygen, moisture and light barrier.

Medium Performance

This type of flexible packaging is typically used for products such as bacon, cheese and boil in the bag foods. It often has an outer layer of Nylon (~20 µm thick), a 4 µm print layer, a 2-3 µm adhesive layer and then a 30 µm polyethylene layer. In common with the general purpose packaging, it has some barrier properties to oxygen, is puncture resistant, thermoformable and can withstand up to 100 °C for 60 minutes.

High Performance

These are typically referred to as 'retort packs' and they have the most complicated construction, which imparts a total barrier, puncture resistance, high heat resistance (e.g., 130 °C for 30 minutes), some thermoforming capability and chemical resistance to the product inside. A typical six layer construction would be:

Polyester outer	18 mm
Print layer	5 mm
Adhesive layer	3-4 mm
Aluminium foil	20 mm
Adhesive layer	4-5 mm
Cast polypropylene (PP)	30 mm

The information in **Table 10** takes into account the great potential for food manufacturers to import reels of the relatively light weight laminated constructions from mainland Europe. In this market there is a 45/50/5 split in the use of solvent-free/solvent-based and water-based adhesive technologies. It is expected with VOC emission regulations becoming more stringent the solvent-free option will grow. For the vast majority of final uses, polyurethane technology is the resin of choice for both solvent-based and solvent-free adhesives. The polyurethane backbone may use polyester or polyether chemistry depending on the quality needed. Water-based adhesives tend to use a combination of dispersible acrylic and epoxy technology crosslinked with amine resins.

The multi-layer construction of flexible packaging always means that there are combinations of adhesives and coatings with printing inks. **Table 14** covers printing ink polymers and additives.

This area is complex, which makes analysis difficult and it is a lot smaller in volume terms compared to coatings. **Table 11** shows that there is a vast split in types of resin technology used, with some of the significant resins which tend to dominate this sector e.g., polyamides and natural rubbers in cold seals, being highlighted. Ethylene vinyl acetate products are used in heat sealable lids.

3.1.4 Inks for Metal Packaging (Table 12)

The bulk of metal decorating inks are applied using lithography and rely on 'paste type inks'. There are four main sectors covered by **Table 12**: beverage, food cans and ends, caps and closures, and general line and aerosols. A brief discussion of each of these is presented next.

3.1.4.1 Beverage Cans

The marketing of all aluminium and tinplate beverage cans relies heavily on a high quality image on the exterior of the thin walled cans. This is the biggest sector for metal decoration and the application is carried out at high speed, 'in the round.' Many combinations of basecoats, inks and over print varnishes are used to provide the right image for a particular product.

The ink technology that is used is based on polyester resins cured with amino crosslinkers. Pigmentation can represent 30% of the formulations, small amounts of diluents (e.g., tridecanol) are used to improve the flow, and lubricants based on PTFE are frequently used to provide the slip to enable post-necking operations to be achieved.

For all the other metal decorating sectors, sheets are printed flat and then post formed.

3.1.4.2 Food Cans and Ends

Few tinplate food cans are decorated in the UK. Some ends are decorated prior to forming. The few can components that are printed tend to use conventional polyester inks without crosslinkers.

In Europe, the emphasis is more on quality and here many cans and ends are printed. It is also the case that the smaller portions packed in drawn cans are distortion printed to provide an accurate and high quality brand image after post-forming.

3.1.4.3 Caps and Closures

The caps and closure markets rely on a printed image on these components to provide differentiation for products packed in glass bottles or jars. Conventional inks are used although there is a growing interest in the use of flexible UV curable technology.

3.1.4.4 General Line and Aerosols

For these sectors there has been a recent move away from conventional ink technology, relying on thermal drying of alkyd technology, to UV curable technology. UV curable inks are now applied to base coated tinplate which is subsequently formed into welded aerosols, milk cans, confectionery tins, and other components.

UV curing technology for flat sheet metal decorating (FSMD) inks use a combination of polyester and epoxy acrylate technologies photo crosslinked using free radical initiators. Cationic UV technology for inks is not used because of the difficulties in achieving a correct ink-water balance.

3.1.5 Inks for Paper and Board Packaging (Table 13)

As with FSMD, UV technology is being used in 80% of the cases for paper and board secondary food packaging. This technology is generally based on free radical curing resins e.g., polyester or epoxy acrylates.

UV curable varnishes are widely used on food cartons and labels.

The conventional inks used in this sector are based on alkyd technology modified with hydrocarbon resins with minor modifications made with lubricants based on PTFE and polyethylene waxes.

3.1.6 Inks for Flexible Packaging (Table 14)

The fastest growing packaging market today uses high quality images produced by reverse printing the exterior clear plastic film layers of flexible packaging constructions. Inks are normally applied using flexographic or gravure high speed printers.

The most widely used resins for these inks are based on a combination of nitrocellulose and polyurethane resins dissolved in fast (i.e., volatile) solvents such as ethyl acetate or isopropyl alcohol. In the UK and Europe, the industry prefers to use low viscosity solvent-based inks, while in the USA water-based acrylic technology is preferred from an environmental and solvent emissions point of view.

Nitrocellulose resins are used to provide excellent dispersion of the pigments in inks. Thermoplastic polyurethanes and other resins are used to plasticise and improve the adhesion characteristics of the formulated inks.

Other resins such as polyamides and polyvinyl butyrals (PVB) have limitations. Polyamides, although providing superior gloss and adhesion properties, are not as effective in heat sealing conditions and suffer from bad odour. PVB are generally used by confectioners as they have low odour characteristics, but only offer below average print flow and heat seal characteristics.

The use of nitrocellulose resins is a safety concern and ink suppliers are trying to replace these by introducing new film forming technology based on high MW polyurethane resins. This approach is seen to be the future as it offers lower migratory technology.

3.2 Harvesting and Processing of Food

The process of harvesting food can bring it into contact with coating products. Some examples are:

Fruit and Vegetables	Picking and transportation to the processing plant.
Meat	Slaughtering and preparation ahead of processing.
Fish	Netting and processing on board. Exposure to painted/varnished surfaces.
Cereals	Coatings on harvesting equipment, milling and storage in bags.
Poultry	Slaughtering, cleaning and preparation for packaging or selling.

Foods are also processed in a wide variety of ways, e.g., sterilisation, drying, chemical preservation, curing and fermenting. Although stainless steel is often the material of choice for the majority of the food contact surfaces in food processing halls, it is possible for food products to, potentially, encounter other coatings, for example:

i) Special acrylic-based coatings on the walls

ii) Polyester or epoxy-based powder coating products on metal substrates

iii) Cold cure, two pack epoxy or polyurethane coatings on the floors

iv) PTFE coated standard conveyor systems

In addition to these, silicone-based coatings have an important role to play, as their versatility means that materials can be manufactured (i.e., epoxy-silanes, silanes, siloxanes, silicones) that can be used to produce hygienic coatings on a variety of substrates such as walls, general surfaces and pipe work. Use of low temperature, UV curing technology can also mean that such coatings can be used on heat sensitive substrates, such as thermoplastics [8]. The performance of these, and other coatings, can be enhanced if quaternary ammonium end groups are present on low MW silicone-based additives, as these confer antimicrobial properties.

This type of development in coating technology is important as improvements to hygiene in food processing plants has become an important topic in recent years [9]. Data published in 2002 by the FAO/WHO Collaborating Centre for Research and Training in Food Hygiene and Zoonoses [10] illustrated the most important places where the food involved in outbreaks became contaminated. This information is summarised below in **Table 15**.

Table 15. Places in the food chain where contamination has taken place	
Place of contamination	Percentage (%)*
Unknown	27.4
Known (listed below)	72.6
Catering**	2.3
Food processing plant	5.4
Kitchen in private homes	6.9
Farm	15.7
Restaurants	38.8
Other	30.9
Sample size: 7,115 **Including hotels, cafes, public houses and bars, and schools*	

Other data in the report includes an assessment of 18,351 foodborne disease outbreaks, of which a contributing factor could be found for 13,310, and 14% of these were thought to relate to improper hygienic conditions that allow contamination during processing. Contaminated surfaces will have contributed to this category.

3.3 Storage and Transportation

Where food needs to be stored in bulk containers, these can be stainless steel, but it is also possible to use coatings to protect more reactive/interactive substrates (e.g., steel and aluminium). For example, the types of speciality coatings that can be used in this type of application include epoxy-phenolic with a glass flake additive, and sprayable elastomeric polysulfide systems. In addition, elastomeric polysulfide sealants, epoxy polyamide holding primers and epoxy grouts can also be used. These types of coatings and coating related products can also be used in other areas of the food and beverage industries.

When food products, such as potable water have to be stored in open topped containers (e.g., in small reservoirs, storage tanks and silos) it is necessary to have an appropriate membrane over the top in order to keep harmful and contaminating agencies out. It is possible to produce such membranes using two–component epoxy-polysulfide liquid coating systems that have Water Regulations Advisory Scheme (WRAS) approval for potable water [11]. These coatings can also be used on a fabric carrier in order to produce a bund type liner.

Food sacks or empty cans are transported on pallets which may be painted or varnished. This gives rise to the possibility that the empty cans could be contaminated by traces of the odourous components used in these coatings. The switch to pallets made from thermoplastic materials eliminates this possibility.

Liquids such as milk are often transported in stainless steel tankers and so no contact with any coatings occurs. However, beer and lager are transported in coated aluminium casks, with this coating usually being spray applied and epoxy/phenolic based.

Within the home, the use of refrigerators to store food is the norm, and there can be problems due to the build up of cold tolerant bacteria such as *Listeria* spp or *Salmonella* spp. It is now possible to treat the surface of a refrigerator with a coating containing a silver glass ceramic antimicrobial additive that has been shown in tests to significantly reduce the build up of *Listeria* [12].

190

3.4 Presentation, Dispensing and Cooking

Packaged products may be exposed to shelving that is polyester or epoxy coated. However, this can be more of an issue with fresh and chilled foods where there is less of a barrier between the food and these coatings.

As with the food processing halls and transportation, the coatings that are used are of the chemical and heat resistant variety to enable fast and efficient cleaning.

Drinks sold in cans are inherently protected from external contamination by the packaging. However, the possibility of contamination of can exteriors from the cleaning chemicals used to clean the surfaces of the vending machine needs to be considered and guarded against.

The use of silicone-based resins to impart non-stick surfaces to cooking equipment such as baking trays, cake tins and loaf tins, and frying pans is increasing, with this technology often replacing PTFE based coatings. These products are being marketed for both use in the domestic environment as well as commercial kitchens. These types of products were included in the recent FSA project undertaken by Rapra [4].

4 Application Techniques for Inks

4.1 Lithography

Lithography is one of the oldest printing techniques. It utilises a flat plate or smooth roller which is treated to accept ink in a selective way. The surface most commonly used is metal, mainly aluminium, although it is possible to use other materials such as ceramics. The surface is treated to provide different degrees of hydrophilicity/hydrophobicity to define the required image. In one form of the process, the plate is then treated with an aqueous solution prior to inking with a solvent-based formulation. In another form, the non-image areas are treated with a silicone-based product before inking (this is referred to as 'waterless lithography'). When the ink is transferred to the substrate for printing via a roller (or train of rollers), the process is called 'offset lithography'.

Conventional lithography has sequential treatments of aqueous and different coloured ink solutions (and possibly a varnish to complete), and successful ink formulations must be resistant to emulsification (which would result in ink bleed to the non-image areas). Lithographic inks are usually of higher viscosity than for other processes and quite commonly are in a paste form. Traditional paste inks are oil-based and use oxidative curing. Shear plays a part in the ink-transfer process and the solvents in litho inks must be less volatile than in other inks to avoid drying on the rollers. Typical solvents in oil-based litho inks are petroleum distillates in the boiling range 260-320 °C or, more recently, long-chain esters derived from vegetable oils, by processes such as alcoholysis, for example:

glyceryl oleate + methanol → glycerol + methyl oleate

The over-print varnish (usually un-pigmented) may be of the same form as the ink, if the application area requires the definition achieved by lithography or may be of the lithographic type (see next) if the application is by a simple roller.

The use of low volatility solvents inhibits drying and so a post-printing heating step is often used. When used with heat-resistant substrates, this heat step can also be used to ensure that the ink is fully cured. On more heat-sensitive surfaces, e.g., paper, absorption of the solvent plays a part in the drying process. In the case of UV-curing inks, more than one irradiation step may be used in the printing stage – for example, immediately after the application of a colour where there is risk of smudging when the next colour is applied.

UV-curing ink formulations for lithography may utilise only low levels of acrylic monomers and rely on multi-functional acrylic polymers to provide the necessary consistency and cure activity.

Lithography offers some of the finest definitions seen in printed packaging and is associated with some of the highest quality products.

4.2 Flexography

Flexography is a relief printing process. A mirrored master of the required image is created in relief on a flexible polymeric printing plate. Flexography uses liquid inks - which can be solvent-based, water-based or UV-curing formulations. It is also capable of operation on a range of substrates, including rough ones such as corrugated cardboard.

The type of solvent that is chosen for the ink must be compatible with the polymer used in the printing plate. For this reason, aggressive aromatic solvents are not usually acceptable, and commonly used ink formulations usually employ alcohol or ester solvents to ensure that a wide range of resins types can be used for the process. In addition, the high solvent levels in liquid inks means that their volatilisation makes a significant contribution to the drying process and ensures that there is less reliance on other drying mechanisms. This in turn means that it is possible to use a wide range of polymer binders in the flexographic inks themselves.

Traditionally, flexographic printing plates were made from vulcanised rubber. The relief is created by vulcanising the rubber, under pressure, against a suitably profiled tooling such as an etched metal mould (e.g., photoengraved magnesium). Solvent resistance is a particular issue with natural rubber, the non-polar synthetic rubbers, and silicone rubber. It is not possible to use inks based on aromatic hydrocarbons with these materials.

Specialty rubbers (e.g., PU, acrylics, fluoroelastomers, and fluorosilicones) offer differing levels of solvent resistance, and liquid moulding/curing technologies (PU and silicone rubbers) provide the scope for moulding against softer tooling (e.g., photo-polymerised plastic). Newly available photopolymers now enable the direct manufacture of resilient flexographic plates without having to use moulding processes, and have given improvements in definition, for better half-tone reproduction, and so on.

Innovations in the materials and technology of plate production are helping to overcome the traditional quality issues associated with compression of the relief at the point of printing, referred to as 'ink squash'. Flexography is emerging as the dominant process for printing on flexible packaging, and is gaining increasing importance on boxes and cartons.

Flexible packaging substrates (i.e., plastic films) are more complex than other substrates in that they are often multilayer to provide an appropriate balance of strength, clarity, permeability and materials' usage.

Printing on flexible film is a high speed process which requires sophisticated mechanisms for tension control. These capabilities, coupled with increasing sophisticated performance requirements mean that lamination and printing are often in-line operations. Lamination after printing, so that the printed surface is sandwiched between two layers, provides so-called 'reverse printing' and dispenses with the need for an overprint varnish.

4.3 Gravure

Gravure is a relief printing process where the image areas are recessed on the printing plate, in contrast to flexography, where the image area stands proud. The plate is made from metal, usually made by chemically-etching copper plated steel. The etched-copper surface is then chromium plated for improved hardness and service life. Gravure provides for high definition printing and requires low viscosity inks for penetration of the smallest recesses in the plate. The use of low molecular weight, i.e., volatile, solvents allows for rapid drying by evaporation, and gravure can achieve very fast printing speeds. However, the high cost of plate manufacture means that gravure is usually restricted to long print runs.

Ink formulations for gravure are selected on the basis of their flow characteristics and the substrate for printing, ensuring that no compatibility issues arise with respect to the printing plate material.

4.4 Inkjet

The mechanics of inkjet printing (nozzle size, etc.) require low viscosity formulations which are mostly dye rather than pigment based. Where pigments are used, they must be exceptionally well dispersed (i.e., particle sizes of 0.5 μm or less). In order to maintain suitably low viscosities without problematic evaporation losses (blocking nozzles, etc.), water-based inks are generally preferred.

Inkjet printing is applicable to a wide range of substrates, and is widely used for printing 'sell by'/'best before' dates and bar codes.

4.5 Influence of Substrate Type

4.5.1 Inks for Metal Packaging

Metals do not provide an absorbent surface and hence the only drying mechanisms are by the evaporation of solvent, and curing. Crosslinking technology figures in most inks and coatings for metal substrates. The bulk of metal decorating inks are applied using lithography and rely on paste type inks.

For beverage containers, made of aluminium, much ink technology is based on polyester resins, cured with amino resins, with a water-based overprint varnish, based on polyester or polyester/acrylic technology. On mainland Europe, the strong steel industry leads to a higher proportion of beverage containers from tinplate. Inks and coatings for tinplate differ from those for aluminium not the least because of a need to provide higher lubricity and abrasion resistance for the harder metal. Lubricants based on PTFE may be used in inks for either tinplate, or aluminium, where useful slip properties are required. Lubricity is important for coated aluminium containers if a shaping operation (e.g., for roll-on screw tops or Monobloc aerosols – see next) is applied after decoration.

For food cans, tinplate predominates mostly using three-piece welded construction. A large proportion of three-piece welded bodies are left uncoated on the exterior as paper labels are used for decoration. A small number of drawn steel 'pie cans' are produced in the UK, which will receive an external combination of white basecoat, inks and varnishes. Direct metal decoration is also more common for food cans from mainland Europe. Conventional inks based on polyester technology are generally used.

Metal caps and closures are encountered on jars and bottles and may be steel or aluminium, with aluminium universal in the ROPP screw top. The latter require very flexible basecoats to withstand the extreme forming process.

Similar flexibility is required for the decoration of aluminium aerosol containers which are used for cream in the UK and other foodstuffs (e.g., ketchup) elsewhere. Aluminium aerosols are commonly made by the Monobloc route where a single aluminium slug is drawn to provide the (one-piece) base and sides of the aerosol can which is usually decorated before shaping the top ('necking') and profiling for the valve seat.

For printing of Monobloc beverage cans, the aluminium profile is supported on a mandrel, and automated fail-safe procedures are put in place to prevent the accidental contamination of the inside of the can which would occur if an empty mandrel was presented to the lithographic plate (i.e., to transfer print to the inside of the next can which arrives on the mandrel). Such accidental transfer is called 'set off', and additional mechanisms for this are discussed later in this section.

Within the UK, a large number of aerosols (e.g., for cream) are still produced by three piece welded construction with tinplate. As with other tinplate decoration, conventional solvent-based polyesters are widely used in the inks and coatings.

For caps and closures, conventional inks are used although there is a growing interest in the use of UV curable technology. The same trend applies in general line and aerosol printing where thermal drying based on alkyd technology is being replaced by UV curable technology.

UV technology for FSMD inks use a combination of polyester and epoxy-acrylate technologies photo-crosslinked using free radical initiators. Cationic UV technology for inks is not used because of retardation of cure due to pigment interactions.

UV curable inks are now applied to base coated tinplate, which is subsequently formed into welded aerosols, milk cans, confectionery tins, and other components.

Also used for food are the so-called 'general line cans'. These are steel cans of various shapes and sizes used for a variety of products, more typically in non-food (e.g., for paint), but these types of cans are used for the bulk packaging of foodstuffs, or where additional presentation is required. Examples include: decorated confectionary or biscuit boxes, decorated boxes or cylinders for premium bottles of spirits and baby food powders. For these products, there has been a considerable move away from thermally cured coatings to using UV inks and varnishes, mainly due to volatile organic hydrocarbon (VOC) regulations and economics. The coloured basecoats, however, are solvent based and produced from a range of resins but mainly polyesters.

Polymer binders for inks that are used on metal substrates are summarised in **Table 12**.

4.5.2 Inks for Paper and Board

Drying by absorption (i.e., penetration of the substrate) can play a part in the drying processes that occur when inks are used on paper and board. This allows for the use of more mobile (i.e., less viscous) ink formulations than with metal substrates. This can compensate somewhat for the limited scope for heat-activated curing, although many papers and boards have coated surfaces, which necessarily allow for rapid uptake. Set off may be an issue unless a speedy chemical cure can be incorporated.

As with flat sheet metal decorating, UV curing is used in the majority of the cases for food packaging involving paper and board. UV technology is generally based on free radical curing resins, e.g., polyester or epoxy acrylates. UV curable varnishes are also widely used on food cartons and labels.

Conventional inks are based on alkyd technology modified with hydrocarbon resins. Minor modifications are made with lubricants based on PTFE and polyethylene waxes.

Polymer binders for inks used with paper and board products are summarised in **Table 13**.

4.5.3 Inks for Flexible Plastic Packaging

The substrates for plastic packaging include various laminated films, whether all-polymer, or laminates which also include foil, paper or metalised layers. Thermal or adhesive lamination may be used. PU is often the resin of choice for both solvent-based and solvent-free adhesives. The PU backbone may use polyester or polyether chemistry depending on the quality needed. Water-based adhesives tend to use a combination of dispersible acrylic and epoxy technology crosslinked with amine resins.

This fast growing packaging market creates for high quality images by 'reverse printing' the exterior clear plastic film layers of flexible packaging constructions.

Inks are normally applied using flexographic or gravure high speed printers. The most widely used resins for these inks are based on a combination of nitro-cellulose and PU resins dissolved in fast-drying solvents such as ethyl acetate or isopropanol. Nitrocellulose resins are used to provide excellent dispersion of the pigments in inks. Thermoplastics PU and other resins are used to plasticise and improve the adhesion characteristics of the formulated inks.

Some limited usage applies to other resins such as polyamides and polyvinyl butyrals (PVB). Polyamides, although providing superior gloss and adhesion properties, are not as effective in heat sealing conditions and suffer through bad odour. PVB are generally used by confectioners as they have low odour characteristics but only offer below average print flow and heat seal characteristics.

The use of nitrocellulose resins is a safety concern and ink suppliers are trying to replace these by introducing new film forming technology based on high molecular weight PU resins. This approach is seen to be promising for the future as it offers lower migratory technology.

Polymer binders that are used for inks in flexible plastic packaging applications are summarised in **Table 14**.

4.5.4 Set Off

The term 'set off' applies to the unintentional transfer of inks and coatings substances from the decorated outer surface of packaging to the inner, food contact surface. The accidental printing of an empty mandrel is one such mechanism of set off. Other mechanisms include the transfer of material when printed sheets are stacked, or are tightly rolled or, in the case of printed paper or board, possible diffusion through the substrate. Although of greater barrier performance than paper or board, plastic films may not be completely impermeable.

Transfer in stacked or rolled sheet is more likely with poorly adhering decoration or if the cure is incomplete. Such deficiencies point to batch or process problems which can impact on all the material processed at the time. On this basis, such set off is unlikely to be missed by QC checks and procedures. The quick and effective detection of set off is usually a stringent requirement of the manufacturers' customers.

A non-destructive test method has been developed by the workers at the Central Services Laboratory, DEFRA, to measure the invisible set off of inks and lacquer on the food contact surface of food packaging [13].

5 Regulations Covering the Use of Inks and Coatings with Food

There are a number of regulatory documents that can be used to demonstrate the safety of coatings and inks intended for food use. A recent review of the regulations for food packaging that exist in both the Europe and the USA has been published by Rapra [14]. In addition, a recent overview of the legislation and testing of food contact materials, as it applies to additives, was given by Sidwell at the 2006 Addcon Conference in Cologne [15]. Each of the principal regulatory documents that are relevant to coatings and inks are briefly described in the relevant sections next.

5.1 Regulation in the European Union

There are no specific harmonised regulations in the EU that cover the use of either coatings or inks with food. However, the Framework Regulation 1935/2004 applies to all materials intended for food contact and so it must be complied with by all manufacturers of printing inks and coatings. This document states that food contact materials and articles:

a) Shall be manufactured in compliance with good manufacturing practice

b) Shall not transfer their constituents to foodstuffs in quantities which could endanger human health, and

c) Shall not bring about an unacceptable change in the composition of the foodstuff or a deterioration in its organoleptic characteristics.

In an annex to the regulation, printing inks are listed among the substances that may be covered by specific measures, although at the time of writing these measures will be far in the future as it is thought that no work on printing inks is in progress in the EU.

In addition to the Framework Regulation, manufacturers must also comply with specific substance documents such as:

i)   BADGE/BFDGE/NOGE Regulation 1895/2005/EC*

ii) Vinyl Chloride Monomer Directive 78/142/EEC

*where: BFDGE = bis(hydroxyphenyl)methane bis(2,3-epoxypropyl)ethers
          NOGE = novolal glycidyl ethers

Regulation 1895/2005/EC, covering food contact materials, articles, surface coatings and adhesives, took into account new toxicological data and placed restrictions on the use of these epoxy derivatives. For example, the use of BFDGE and NOGE were prohibited as from 1st January 2005 and 1st March 2003, respectively, with the exception of heavy duty coatings in tanks having a capacity greater than 10,000 litres. For BADGE, and its hydrolysis products, a migration limit of 9 mg/kg of food was set, and for BADGE chlorohydrins the limit is 1 mg/kg of food. This legislation is of historical interest with the EU as it was the first to set out any rules that are specific to food contact coatings.

One EU document that does refer specifically to coatings and inks is the Regenerated Cellulose Film Directive – Directive 93/10/EEC (as amended by Directives 93/111/EEC and 2004/14/EC). The positive list in this regulation does not include dyes, pigments and adhesives, and substances used for these purposes are not allowed to migrate into food in detectable amounts. If a plastic coating is to be applied to the film, only substances in the lists of authorised substances in the Plastics Directive 2002/72/EC (as amended) can be used and the whole film has to comply with the overall and specific migration limits laid down in this Directive.

There is in existence a draft version of a 'Super-regulation' for plastics which includes products that are comprised of multi-material layers on the condition that the food contact surface is a plastic. The definition of these products is such that the print and the print substances could be regarded as a layer. This document also specifies that printing ink, when used to manufacture articles for food contact, should be included in the determination of overall migration.

It is also worth mentioning in this section, the Synoptic Documents that are issued by the EU. These summarise the status of the substances listed and give the current evaluations by the Scientific Committee for Food (SCF). The Synoptic Document that is of relevance here is entitled *'Draft of provisional list of monomers and additives used in the manufacture of plastics and coatings intended to come into contact with foodstuffs'*. Although it has been in existence for a number of years, this working document is a provisional and incomplete list of all the monomers and additives that are used for food contact plastics and coatings (excluding silicones) within the member states of the EU. It is a working document and is not legally binding, and updated lists are issued as different substances are evaluated by the SCF. These substances are submitted by industry, often through trade organisations, and are then examined by the SCF from a toxicological viewpoint and classified into one of ten lists – numbered 0 to 9. There is also a List W (Waiting List) for substances that are not yet included in the positive lists of the member states. Although these substances appear in the Synoptic Document they cannot be included in the EU list as they lack the data required by the SCF.

5.2 Council of Europe (CoE) Regulations

5.2.1 Coatings

There is a *CoE Resolution on coatings intended to come into contact with foodstuffs* – Resolution ResAP (2004). This document covers the following types of food contact coating:

a) Coatings for metal packaging

b) Flexible packaging coatings

c) Heavy duty coatings

A 'coating' being defined by the Resolution as 'the finished material prepared mainly from organic materials applied to form a layer/film on a substrate in such a way as to create a protective layer and/or impart technical performance'.

In addition to the Framework document, this Resolution has two Technical documents associated with it:

1. Technical document No.1. Contains an inventory of substances to be used in the manufacture of coatings intended to come into contact with foodstuffs. This list has been compiled with the help of The European Council of Paint, Printing Inks and Artist's Colours Industry (CEPE).

2. Technical document No. 2. This document contains the scientific background for the elaboration of the Resolution

The Inventory is divided into two categories, monomers and additives, all of which have some national member state or FDA authorisation. Both of these sub-lists are divided into those which are already fully evaluated and included in SCF/ European Food Safety Authority (EFSA) lists, and those that have not yet been fully evaluated by SCF/EFSA (called the Temporary Appendix). The Temporary Appendices to the monomer and additive lists are time limited with a deadline for evaluation of five years from the date that the Resolution is adopted. Although this Resolution, in common with other CoE Resolutions, has no legal status, it is regarded as being useful in demonstrating compliance until such time as the EU has fully incorporated surface coatings into its legislative framework. In addition to being manufactured according to 'good manufacturing practice' (GMP) using the monomers and additives listed in the Inventory, the Resolution states that the Resolution on polymerisation aids – AP (92)2 should also be adhered to. The Resolution also stipulates an overall migration limit of 10 mg/dm^2, or 60 mg/kg of food for certain specified situations, as measured by EU methods (see Section 6.1).

The Resolution also states that coatings should not transfer migrating components, not listed in Technical document No. 1, which have a molecular weight of less than 1000 daltons in quantities which could endanger human health. If such migrants are detected, they should be subjected to appropriate risk assessment taking into account dietary exposure as well as toxicological and structure-activity considerations (see Section 6.6).

For those coatings that are based on silicone type polymers, there is a specific CoE Resolution, APRes (2004). Both silicone rubbers and silicone resins are covered by this Resolution. Blends of silicone rubber with organic polymers (EPDM rubber) are also covered by the Resolution provided that the silicone monomer units are the predominant species by weight. There is an overall migration limit of 10 mg/dm^2 of the surface area of the product or material, or 60 mg/kg of food. There are restrictions on the types of monomer that can be used to produce the silicone polymers and there is an inventory list : *'Technical document No.1 – List of substances used in the manufacture of silicone used for food contact applications.*

Where rubber type protective coatings are used, there is no specific EU legislation, but there is a CoE Resolution, APRes (2004). The Resolution contains an inventory of additives (Technical document No. 1) and a small section on breakdown products – nitrosamines and amines. In addition, there are four other technical documents, covering such things as good manufacturing practice, and test conditions and methods. The Resolution also places rubber products into one of three categories according to their application and the migration that may occur. Migration testing is required for only the first two of these, where an overall migration limit of 60 mg/kg food applies.

5.2.2 Inks

The CoE has recently issued a Resolution on *'Packaging inks applied to the non-food contact surface of materials and articles intended to come into contact with foodstuffs'* – Resolution AP(2005)2. As the title suggests, inks that come into direct contact with food are not covered by this Resolution. Also excluded are inks used on the outside of glass bottles and metal cans (as the substrate material is regarded as being a comprehensive barrier). Its main purpose is to regulate the inks used on the outside of plastic, and paper and board type food packaging. One of the problems that were encountered in the drafting of the resolution was the understandable reluctance of industry to release detailed information on the formulation of ink products. The Resolution states that the ink supplier is responsible for the composition of the ink. There are three technical documents that accompany the Resolution:

1) Technical document No. 1: Contains an inventory of substances being used in the industry, an exclusion list, and specific migration limits where possible.

2) Technical document No. 2: This is a GMP Guide, in two parts. The first part addresses inks and has been drawn up by CEPE, the second, which addresses the plastic and paper and board substrates has been prepared by the European Forum of Flexible Packaging Industry and the International Confederation of Paper and Board Converters (CITPA).

3) Technical document No. 3: This document provides guidance on the conditions that should be used for the testing of packaging inks applied to the non-food contact surface of food packaging.

The main CoE Ink Resolution requirements can be summarised as follows:

i) The packaging inks should be manufactured and applied on the support in accordance with the guidelines for good manufacturing practice and with the recommendations of the converters as set out in Technical Document No. 2

ii) The printed or overprint varnished layer of finished printed materials or articles should not come into direct contact with food.

iii) There should be no or only negligible visible set-off or migration from the printed or varnished non-food contact layer to the food contact surface.

iv) The overall migration and specific migration level (SML) for the substances in the inventory lists apply.

v) Migrated printing inks shall not be detectable at the lowest concentration at which a substance can be measured with statistical certainty by a validated method of analysis, i.e., a detection limit of 10 ppb or below.

This Resolution has attracted some controversy, with industry groups in the EU (e.g., The European Printing Ink Association (EuPIA) – a sector of CEPE) saying that there will be problems with its implementation, and in some cases that it is unworkable in its present form [16]. There are a number of reasons cited for this, for example:

a) The Inventory is incomplete, with only representative examples of generic groups present, and many key raw materials are missing altogether, e.g., there are no adhesion promoters (used with flexible packaging inks on polyolefin films), no phenolic resins (used in most sheetfed offset litho inks), and no photoinitiators (used in UV-cured inks).

b) The majority of the substances have not been evaluated and so migration studies have to be carried out using a demanding and, in some cases unfeasible, 10 ppb detection limit. The end result could be that some inks become unavailable.

These industry concerns have led to groups such, as the British Coatings Federation (BCF) in the UK, to compile their own, alternative Inventory lists. A draft version of the BCF list (as it stood in June 2006) is included as an Appendix in the final report for the FSA Coatings and Inks project [17]. There are also other industry driven documents such as the EuPIA *Guideline, Declaration of Conformity and Good Manufacturing Practice*. These documents, which are available from the EuPIA web site (*www.eupia.org*) were issued to states within Europe during 2005 and 2006. There is also a EuPIA funded investigation into ink additives that have not been evaluated and listed, and a commitment to reduce the migration limit of 'no concern' for these non-evaluated substances to 50 ppb by 2010, with a further reduction to 10 ppb by 2015, which will bring inks in line with other food contact materials [16].

5.3 National Regulations within the EU

The regulations that exist within certain states in the EU can be used to demonstrate compliance with the Framework Regulation. For coatings, the most comprehensive of these is the Dutch *Verpakkingen-en Gebruiksartikelenbesluit (Hoofdstuk X)*. This contains a positive list of permitted starting substances and is widely used within the EU. In those cases where thermoplastic polymers are used as the coating (e.g., on a metal substrate), the plastic food contact materials and articles Directive 2002/72/EC can often be used to demonstrate food safety. However, food use coatings are far more likely to be of the thermosetting type (see Section 2.1), which are more complex and not fully covered by 2002/72/EC. In these cases, reference to the CoE *Resolution on Surface Coatings* (see Section 5.2), or the published opinions of the Scientific Committee on Food/European Food Safety Authority (SCF/EFSA) can be used to demonstrate compliance.

In Germany, *Recommendation XV of the BfR regulations* covers silicone rubbers and silicone resins – both of which can be used as coatings. The relevant sections stipulate acceptable starting materials and additives that may be used in processing and manufacture – both types and maximum levels. There are overall limits for volatile organic material as well as total extractable material, and specific limits for certain starting substances, e.g., residual peroxides in the case of silicone rubbers.

In the case of rubber coatings, there are a number of national regulations that can apply. The most prominent of these include [18]:

a) French requirements - Arrete of November 9th 1994, Journal Official de la Republique Francaise, 1994.

b) Dutch regulations - Verpakkingen en gebruiksartikelenbesluit (Chapter III). These are similar to the CoE Rubber Resolution.

c) Italian requirements - Supplemento ordinario alla Gazetta Ufficiale della Repubblica Italiana, 1973.

d) United Kingdom requirements - *Statutory Instrument 1987 No 1523 Materials and Articles in Contact with Foodstuffs.*

The French and Italian regulations cited previously are the general food contact documents and so other coatings materials, in addition to rubber, are also covered by them. Food contact materials in Belgium are regulated under the Royal Arrete of May 1992 on Materials Intended for Contact with Foodstuffs. A total of seven EU Member States (Belgium, France, Germany, Italy, the Netherlands, Spain the UK,) have some form of National positive list of permissible substances for use in manufacturing food contact materials in addition to the EU Directives. A listing of national food packaging legislation is available from the European Commission and additional information is available in book written by Ashby, Cooper, Harvey and Tice [19].

With respect to inks, there a very few National regulations within Europe. An exception is Chapter XXXVI of the German *Recommendation BfR*, which has a general statement on colorants and optical brighteners in the food contact material. This states that they should not migrate into foodstuff and that no testing is required for packages intended for dry, non-fatty food.

5.4 FDA Regulations

The food regulations for polymer and resinous coatings are given in the FDA *Code of Federal Regulations (CFR) 21.*These documents, which are used globally for the formulation of a large number of coating products, list authorised starting substances and lay down test conditions and migration limits. Coatings for specific substrates can be found in the following sections of CFR 21:

Metal substrates and repeated use	175.300
Polyolefin films	175.320
Paper coatings	176.170

The current regulations in the US for both coatings and adhesives have been reviewed by Baughan [20]. At the same PLACE conference, Lin and co-workers [21] presented a paper describing developments and improvements that had been made to the migration tests that are necessary to establish if electron beam (EB) and UV cured coatings and adhesives complied with the FDA regulations. The contribution of relatively new analytical techniques, such as liquid chromatography (LC), mass spectrometry (MS)/MS, to this type of work is featured.

Inks are very rarely used in direct contact with food because there are virtually no ink formulations that comply with the FDA. For example, there a very few carbon black pigments which are acceptable to the FDA and, to achieve coloured inks, only food colorants are permissible. This latter requirement severely limits the methods that can be used for the application of the ink. In addition, manufacturers of food contact inks have to ensure that no carry over was possible from equipment that had been used in the manufacture of non-contact products (e.g., for external decoration on cans and metal tins etc), which are not required to be FDA compliant.

5.5 Other Considerations for Industrial Use

In addition to the Regulations described in Sections 5.1 to 5.4, there are other reference documents that industrialists need to be aware of:

a) REACH EU Regulation

    This Regulation covers the Registration, Evaluation and Authorisation of Chemicals (REACH) of chemicals and might have an impact on the substances used in coatings and inks in the future. The cost of compliance may mean that some small volume chemicals disappear from the market place, although implementation may take over 10 years for some substances. Importantly, as polymers are exempt from the Regulation those used in the manufacture of coatings and inks will not need to be registered.

b) CEPE Exclusion list for Printing Inks

    In addition to its GMP guides, CEPE has published an exclusion list for printing inks. The substances on this list are present in the *Dangerous Substances Directive* (67/548/EEC), and pigments colorants based on antimony, arsenic, cadmium, chromium(VI), lead, mercury and selenium. Some solvents are also on the list, as are cyclohexane, and some stilbenes, butylphenols and benzophenones compounds. This CEPE list is not identical to the CoE exclusion list.

c) British Coating Federation Documents

    The BCF has issued a guide to printing inks for food wrappers and packagings. At the time of writing it is also working on an inventory designed to be an alternative to that contained within the CoE Resolution on *Inks – Technical document No. 1* (Section 5.2). The reason that the BCF believes that an alternative list is necessary, is that whereas the CoE Inventory list mainly originated from national legislation their list relates to those substances currently used in Industry. To help understand the regulatory status of the BCF listed substances, the two lists have been compared in the final report of the Rapra FSA project (A03055) [17] and those substances not in the COE Inventory list have been highlighted. This exercise showed that a relatively large number of substances in the BCF list are not included in the CoE list.

d) Good Manufacturing Practice

The coatings and ink industries have their GMP guidelines to ensure that their products are manufactured to the highest standards.

e) CEPE Code of Practice for Coated Articles where the food contact layer is a coating

This extensive and wide ranging, industry driven, code of practice will be taken by the Commission as a base for a specific regulation on coatings. Work on this document is on-going at the time of writing - Draft 10 was published on 30th June 2006. This code and its guidelines, which has similarities to the Plastics Super Regulation and the 4th amendment to the Plastics Directive 2002/72/EC (see Section 5.1), and which describes how compliance with the Framework Regulation 1935/2004/EC can be demonstrated for direct food contact coatings, are voluntary in nature and individual companies may decide to apply them either fully or partly, according to their own judgement. A list of the companies supporting the code can be obtained from CEPE.

The Code of Practice applies to the food contact surfaces of the following:

i)   Coated light metal packaging up to a volume of 10 litres.

ii) Coated metal pails and drums with volumes ranging from 10 to 250 litres.

iii) Coated articles with volumes 250 to 10,000 litres.

iv) Heavy duty coated articles having a volume >10,000 litres.

v)   Coated flexible aluminium packaging.

Sectors which will be incorporated at a later date when more details are available are:

i)   Printing inks in direct food contact.

ii) Coated primarily used to seal food packaging.

iii) Gaskets for metal closures.

iv) Coated flexible plastic packaging.

v)   Coatings and inks for paper and board.

The Code of Practice does not apply to:

i)   Repeated use, non-stick coatings, which remain regulated by the specific chapters of BfR, VGB, and FDA applicable to them.

ii)   Extrusion coated materials or articles where the extrusion coating being a plastic, should comply with the provisions of Directive 2002/72/EC, as amended.

iii) Laminated packaging articles or components where the food contact layer, being a plastic, should comply with the provisions of 2002/72/EC, as amended.

iv)   Printing inks and coatings applied to the non-food contact surface of food packaging materials and articles intended to come into contact with foodstuffs.

v)   Coatings on paper and board which remain regulated by specific chapters of BfR, VGB and FDA applicable to them.

vi)   Coatings on regenerated cellulose which are covered under Commission Directive 93/10/EEC and its amendments.

vii) Can end sealants based upon rubbers and elastomers which remain covered by rules applicable under national legislation.

viii) Tin coatings, wax coatings and adhesives

6 Assessing the Safety of Inks and Coatings for Food Applications

As this review illustrates, many different types of polymers and types of polymer product (e.g., thermoplastic, thermoset and rubber) can potentially be used for coatings in food applications (see Sections 2.1.1 to 2.1.11).This in turn means that a large number of different regulations and requirements may have to be addressed, in order to decide upon the correct conditions (e.g., choice of simulant, test samples, times and temperatures) under which to carry out food migration testing. This point is illustrated by the summary of regulations which is provided in Section 5.

For direct food contact coatings, the CEPE Code of Practice (see Section 5.5) provides a great deal of useful information on how compliance with the Framework Regulation 1935/2004/EC can be demonstrated. An overall migration limit of 60 mg/kg is provided, and SML for particular substances are provided in the Annexes. In addition, guidance on the food simulants for the migration testing is provided, as are guidelines on assessing exposure to migrating substances and basic rules for demonstrating compliance with the overall migration limit (OML) and the SML.

Inks for food use, on the other hand, do not present such a complicated problem as they represent a specific product category in themselves, are mainly used on the non-food contact service of food packaging, and the binders in them do not span such a wide range of polymers types and chemistries.

There is an on-going effort to ensure that migration data is as accurate as possible. For example, the EU recently funded a research project (EU AIR 94-1025) to facilitate the introduction of migration control into GMP and into enforcement policies and a part of this project involved the re-evaluation of the analytical approaches to extract and identify potential migrants from food contact materials. The results of this part of the project are published in the journal, *Packaging Technology and Science* [22] and the journal *Food Additives and Contaminants* [23].

There is also a continuing need to detect migrating substances at lower levels. For example, toxicologists have suggested that species which are ingested in amounts exceeding 1.5 mg/day should be identified and toxicologically evaluated. Identification and quantification of species at such a low level will cause problems and these are discussed in a paper by Grob [24]. The conclusion reached is that it will be difficult to achieve the comprehensive analysis of coating migrants down to concentrations that are presently regarded as safe with the analytical instrumentation currently available.

6.1 Global Migration Tests

This is the usually the simplest test to perform and it is used to determine if the product is suitable for a particular food use application by passing the global migration limit that is stipulated by all of the various regulations covered in Section 5.

The methodology of the test varies depending on the regulation that is being addressed, as does the way of expressing the data and the limits that have to be met. Specific details can be obtained from the various pieces of legislation.

For example, in the EU, the global migration tests that apply to coatings and inks, and other food contact materials, are described in the following documents:

i)   Council Directive 82/711/EEC of 18th October 1982, as amended by Commission Directive 93/8/EEC of 15th March 1993

ii) Commission Directive 97/48/EC of 29th July 1997

iii) Council Directive 85/572/EEC of 19th December 1985

The food simulants that are specified in these Directives are:

a) Aqueous foods - Distilled water

b) Acidic foods – 3% acetic acid

c) Alcoholic food – 10% ethanol

d) Fatty food – olive oil

Migration testing using these simulants should be performed under the worst foreseeable contact times and temperatures that can be envisaged for the application. For example, long-term storage at room temperature is represented by testing for ten days at 40 °C. The 82/711/EEC document provides a correlation table for migration test conditions. The analytical methods for testing overall, and specific migration, have been standardised at the European level by CEN (the European standardisation body).

6.2 Specific Migration Tests

These tests are used to target specific chemical compounds for which there is a toxicological concern and a SML, i.e., listed substances. In common with the global migration test, the tests specified (target species and test conditions) vary from regulation to regulation, but some species appear regularly due to the degree of concern associated with them.

There are a number of cases where there are specific analytical test methods documented, particularly in cases where there are resolutions or regulations on a particular migrant. For example:

i)   Free vinyl chloride monomer – Analytical methods described in Commission Directives 80/766/EEC and 81/432/EEC

ii) Determination of 4-methyl-1-pentene in food simulants – CEN/TS 13130-25:2005 [25]

iii) Determination of bisphenol A in food simulants – CEN/TS 13130-13:2005 [26]

iv) Determination of 1-octene and tetrahydrofuran in food simulants – CEN/TS 13130-26:2005 [27]

The test methods that should be used for specific migrants from coatings that can be regarded as plastics are described in EN 13130-1 [28].

There are also cases where a particular species has attracted a lot of analytical attention in recent years. A good example of this is BADGE from can coatings and a number of approaches for its determination have been documented [29].

6.3 Fingerprinting of Potential Migrants from Coatings and Inks

It is often useful to produce a qualitative or semi-quantitative fingerprint of the low molecular weight species in coatings and ink products that have the potential to migrate into food. Gas chromatography-mass spectroscopy (GC-MS) is often used for this due to its high resolution and the identification power of the mass spectrometer. In order for the data to be representative, it is important that the coating and ink product is in the form in which it is used in the final end use product. As both of these materials usually have to undergo a curing or drying stage, this should be carried out prior to any analysis step, or samples taken from the final product *in situ*. The latter option can present a problem with inks when only small amounts of material are used, for example., in printing. The considerations detailed previously will mean that the samples are in the solid phase, and so the headspace GC-MS techniques are the most applicable, with the dynamic version having the advantage over the static in that it requires less sample. The solvent extraction GC-MS technique may be applicable where larger areas of coating and print are available for analysis. This method has the disadvantage, however, of the initial solvent front obscuring early eluting (i.e., low molecular weight) species. The relatively recent commercialisation of two dimensional GC-MS instruments has provided the analyst with greater resolving power, coupled with improved detection limits and enhanced deconvolution software [30].

As in-house coatings and inks specific databases are developed for LC-MS, the inclusion of this technique into the fingerprinting process will complement GC-MS data by contributing information on thermally labile, relatively large (e.g., oligomers), and highly polar (e.g., organic salts) potential migrants.

Albert and co-workers [21] have reviewed, and compared, the use of modern liquid chromatography methods, such as LC-MS and LC-MS/MS, with the GC-MS technique for the analysis of potential migrants from FDA compliant, EB and UV curable food packaging coatings and adhesives. In addition to comparing the analytical capability of these systems, the effect that various processing variables (e.g., curing voltage and dosage) had on the extractable data were also evaluated.

6.4 Determination of Specific Target Species in Coatings and Ink Products and in Food Simulants and Foods

The use of specified tests to determine species that have SML has been covered in Section 6.2. There are reasons why further analytical testing is often required, for example to ensure that a coating or ink has been formulated using only additives and ingredients that are present in a particular positive list, or to quantify a potential migrant for which there is as yet no specific migration test or SML. A review of the analytical methods that can be used to identify and quantify a range of species that can be found in food contact materials has been written by Veraart and Coulier [31].

For convenience, species have been placed into functional categories next and the analytical techniques used to detect and quantify them explained.

6.4.1 Monomers, Solvents and Low Molecular Weight Additives and Breakdown Products

Monomers are either gaseous or relatively volatile liquids and so GC and GC-MS based techniques are usually used to determine them in both the final coating and ink product and the food simulant/food product. To simplify the analysis, a static headspace, or dynamic headspace, sampler is often used to isolate the analyte from the sample matrix [32], an extraction procedure often presenting problems due to the masking effect of the solvent. There are important examples, however, where analysis, sometimes following chemical derivatisation work is used. This approach was used by Paz-Abuin and co-workers [33] to carry out specific migration work on two epoxy resin coatings for use with drinking water. The amine curing agents were determined by high performance liquid chromatography (HPLC) with pre-column derivatisation, whereas the epoxy resin monomer was quantified using HPLC without any derivatisation. A reverse phase HPLC system with fluorescence detection was used in both cases.

In addition to antidegradants and curatives (which are mentioned next), other additives (e.g., pigments) can also produce breakdown products which are regarded as harmful. For example, Michler's ketone, can be a degradation product of certain violet dyes used in printing. This compound is regarded as a potential carcinogen and so methods to assess its migratory behaviour into food simulants have been developed [34].

The use of GC-MS to determine the levels of photo-initiators and acrylic esters which have the potential to migrate from inks into food simulants has been described by Papilloud and Baudraz [35]. The study covered the migratory behaviour of nine different acrylate monomers and six different photoinitiators in a range of aqueous and fatty food simulants.

6.4.2 Oligomers

Prior to the commercialisation of LC-MS instruments, supercritical fluid chromatography was widely used for the analysis of oligomers. As the molecular weight range of LC-MS instruments can be extended up to 4000 daltons this capability makes them ideal for the detection and quantification of oligomers. For example, it has been shown that silicone oligomers can be detected by LC-MS in food simulants [36], and [4].

6.4.3 Plasticisers and Oil-type Additives

These additives are essentially high boiling point liquids and so the most appropriate technique to use is LC-MS. A range of synthetic plasticisers such as phthalates, adipates, mellitates and sebacates can be detected using the atmospheric pressure chemical ionisation mode. If data on non-polar hydrocarbon oils is required, then an atmospheric pressure photoionisation head (which can detect non-polar species is required) or, if the oil has a sufficiently high aromatic character, in-line UV or fluorescence detectors can be used.

6.4.4 Polar Additives and Metal Containing Compounds

For additives that are highly polar (salts or ionic compounds), for example, antistats, thickening agents and surfactants, there are two analytical techniques which can be of use to the analyst: an LC-MS fitted with an electrospray head, and anion and cation ion-chromatography. Both of these have the potential drawback that they are much easier to use on aqueous samples, rather than fatty ones- an intermediate extraction step often being required in the latter case.

In certain cases, compounds which have a metal component (e.g., platinum catalysts used in certain silicone products) are present in the coating or ink product. In order to determine these types of migrants at a low (i.e., ppm) level, techniques such as atomic absorption spectrophotometry and integrated coupled plasma have to be used. One complicating factor which needs to be borne in mind with this type of analysis is that a value for the target metal will be obtained irrespective of which compound/ additive it is in; interferences can therefore occur and knowledge of the product's composition and/or the service history of the sample are important.

6.4.5 Cure System Species, Initiators, Catalysts and Their Reaction Products

These species are usually low, or relatively low molecular weight organic compounds of intermediate polarity and as such as ideally suited for determination by GC-MS.

Problems can occur if the species are very thermally labile and/or reactive, and in these cases (as in the cases of metal salts – see previously) LC-MS is the preferred technique. It is also easier to use LC-MS with a number of the approved food simulants as they can be injected directly into the instrument, being compatible with the mobile phase.

6.4.6 Antidegradants, Stabilisers and Their Reaction Products

This class of additives is generally less thermally labile and reactive than the preceding one and so GC-based methods can be used for a number of them. Where high processing temperatures dictate the use of relatively high molecular weight, oligomeric type stabilisers, LC-MS methods have to be used.

6.5 Sensory Testing

This category covers subjective testing by a panel of trained human assessors to identify taints and odours in food packaging. The number of assessors in the panel is determined by the sensitivity that is required for the test, i.e., the statistical reliability. Although the test is subjective, experience has shown that a well trained panel will produce consistent results.

The transfer of a taint is more important from a legal point of view, whereas an odour has more importance from a marketing point of view.

For the evaluation of the odour of printed paper, board, plastic or any other material, test pieces are stored in jars for a certain time. The odour present in the jar air is then assessed by the panellists and the intensity rated on a scale such as 0 (no odour) to 4 (strong odour). To evaluate taint, test pieces are stored with a test food (e.g., chocolate) in a jar and then the taint of the food is assessed and rated by the experts. There are a number of ways of performing this operation. For example, in a 'triangle test', one portion of food has been stored with the material to be tested, the other two are reference samples, and the assessors have to select the odd one out. In a 'multi-comparison test', the assessors are given a reference food sample that is regarded as having a value of 0 and then they score the intensity of the taint of an analysis sample using a scale usually of 0 (no taint – i.e., same as reference) to 4 (strong taint).

There are no specific international standards for assessing odour and taint as a result of inks and coatings, but there is a general ISO standard, ISO 13302:2003 [37] that addresses the modification of the flavour of foodstuffs due to packaging materials. There is also a specific standard for paper and board, EN 1230-1 [38] and two draft standards on sensory analysis in general, ISO 4120:2004 [39] and ISO 5492:2005 [40].

Because carrying out sensory testing for odour by a human panel is time consuming, and of increasing importance in a number of other manufacturing sectors (e.g., automotive), a great deal of effort has been expended in the search for a device that can act an effective and reliable 'electronic nose'. It now seems that such an instrument, which is based on a chemical sensor array, will soon be acceptable for routine analysis within the food and packaging industry. An investigation by van Deventer [41] on volatile chemicals from inks on plastic films showed that the quartz sensors of an electronic nose system were able to discriminate between packages with different levels of retained solvents, and Frank and co-workers [42] demonstrated that an instrument containing eight sensors and eight microbalances gave a good correlation with human assessment panels concerning retained solvents in printed wrapping foils.

6.6 Toxicological Assessment of Migrants

Defining the starting substances for coatings and inks can depend upon which part of the industry a person comes from – the ingredients industry or coatings manufacturing. The list that results (monomers, additives, polymerisation aids, resins, oligomers, additives) is therefore very extensive.

This large list, coupled with the complex chemistry associated with the manufacture of the starting substances and the chemical reactions that take place to produce the finished products, leads to a very complex mixture of substances having the potential to migrate into food simulants and food. A CoE working group has shown that it is impossible to evaluate all of these substances from the toxicological point of view according to EU guidelines.

One way of dealing with this problem for substances that are not listed (i.e., their toxicity is not known), is to use a 'tiered' approach, which takes into account dietary exposure as well as toxicological and structure-activity considerations, the aim being to reduce the number of toxicological evaluations that are required.

For example:

1) Tier One

Substances which migrate into food or food simulant at less than 10 ppb should not be evaluated.

2) Tier Two

Many substances used in coatings can be reduced to smaller units (e.g., the oligomers and polymers to the monomers). As this smaller unit represents the 'worse case scenario' in terms of toxicity it is toxicologically assessed and the result applied to the oligomers and so on.

In accordance with EU guidelines of toxicity testing, the following applies:

i) >10 ppb up to < 50 ppb   Three mutagenic tests

ii) >50 ppb up to < 5ppm   Three mutagenic tests,
                              90 days feeding study

iii) >5 ppm   Three mutagenic tests, 90 days feeding study,
                              2 years feeding study

3) Tier 3

The decision to use Tier 2 is based on state of the art knowledge and/or QSAR. If any scientific knowledge and/or QSAR show that toxicity of a monomer is not equal or less than the toxicity of the oligomer/polymer, an evaluation of the oligomer/polymer has to take place using the same toxicity guidelines as given for Tier 2.

Note: QSAR = Quantitative Structure-Activity Relationship. A mathematical model that relates a quantitative measure of chemical structure (e.g., a physicochemical property) to a property or to a biological effect (e.g., a toxicological endpoint). QSAR are being used more extensively to save time, money and in the interests of animal welfare.

7 Potential Migrants and Published Migration Data

7.1 Acrylates

The lower acrylates are known irritants, and ethyl acrylate is regarded by the IARC as a Group 2B carcinogen. However, there are many acrylate monomers used in ink and coating formulations and these lower acrylates may be far from representative. Of higher molecular weight, and commonly used in UV-curing inks are the acrylate esters of polyhydric alcohols. These types of acrylates have been detected and quantified by GC-MS [35]. Reference materials having molecular weights up to 470 were determined with a detection limit of 20 ng. The quantification of these monomers in food migration samples can prove difficult owing to the complexity of commercial products (which may be mixtures), and high MS acquisition rates, or the use of two dimensional (i.e., GCxGC) instruments may be necessary to resolve the individual components (see Section 6.4).

7.2 Amines

The aromatic amines can be of particular concern as some of these compounds are classed as carcinogens. Although the most potent carcinogens are found amongst two-ring aromatics (benzidine, naphthylamine), even single ring compounds such as *ortho*-toluidine or *ortho*-anisidine (2-methoxyaniline) are found on the IARC listings. Both of these single ring aromatics have been detected in aqueous extracts of printed multi-layer plastics, and *ortho*-anisidine has also been detected in olive oil extracts [43]. The same studies also found 2,4-dimethylaniline in both aqueous and olive oil extracts. For all three amines, detection in olive oil was by headspace GC-MS (detection limit 20 µg/kg) and, in water, solid-phase extraction was used followed by GC-MS (detection limit 0.1 µg/kg).

A significant amount of research work has been carried out on the amines which are the precursors and hydrolysis products of the isocyanates in PU, notably the three aromatic amines, the 2,4- and 2,6- isomers of toluene diamine (2,4- and 2,6-diaminotoluene), and 4,4′-methylenebis(aniline). Potential sources are the PU used in coatings and ink binders, or in the laminating adhesives used for multilayer packaging. All three amines were detected in aqueous extracts of laminated films using derivatisation followed by GC-MS [44]. Solid-phase derivatisation with trifluoracetic anhydride ($RNH_2 \rightarrow RNHCOCF_3$) was used, and the respective detection limits were in the range 0.1-0.4 µg/l.

Derivatisation is necessary with these types of compound to avoid peak tailing in GC analysis. Other reagents which can be used for this purpose include hexafluorobutyric anhydride, as in the OSHA 1910 method for airborne species, with GC electron capture detection for the subsequent analysis step. Derivatisation can also be performed to enhance sensitivity to UV detection, and this forms the basis for the global analysis of primary aromatic amine by diazotisation and coupling with(1-naphthyl)ethylenediamine dihydrochloride [45] [46]. This method has a detection limit of 8 µg/kg (total aromatic amine quantified with respect to aniline), and provides a valuable screening technique.

A recently reported technique which provides good sensitivity without derivatisation uses electrospray ionisation with tandem mass spectrometry [47]. The technique is rapid as there is no chromatography step and the reported detection limits were 2-3 ng/ml in aqueous ethanol extracts.

One aromatic amine which does not arise specifically from colorants or PU is 4,4′-bis(dimethylamino) benzophenone [$(Me_2NC_6H_4)_2C{=}O$, Michler's ketone]. Michler's ketone is a photoinitiator for UV-curable inks. Although its potential carcinogenicity excludes its direct use in food contact materials, Michler's ketone is a well-established initiator, and concerns have been raised over its introduction into food packaging via recycled fibres [34].

7.3 Aromatics from Unsaturated Polyesters

Amongst the species considered here are: acetophenone, benzaldehyde, benzene, ethylbenzene and styrene. Benzene and ethylbenzene are classified by IARC as Group 1 and Group 2A carcinogens, respectively. Benzaldehyde is harmful by ingestion and has a potential to cause allergic reactions.

Benzene can be found in unsaturated polyester resins due to it being a relatively minor component of the styrene feedstock. Styrene being the most commonly used reactive diluent for these types of resins.

Dynamic headspace GC with flame ionisation or MS detection has been used for the determination of low molecular weight aromatics in cured unsaturated polyesters [48]. At 200 °C, volatilisation of ethylbenzene and benzaldehyde was complete after 1 hour, whilst styrene continued to evolve as a result of on-going depolymerisation. The levels determined (at 200 °C) included: up to 37 mg/kg of ethylbenzene, up to 180 mg/kg of benzaldehyde and up to 1330 mg/kg of styrene.

The dynamic headspace GC, or GC-MS, method is unsuitable for the determination of these aromatics in foodstuffs, or aqueous simulants, because of the interference of water. Gramshaw and Vandenberg extended their studies into foodstuffs using a workup procedure which included azeotropic distillation and pentane extraction to find up to 188 μg/dm² styrene in pork cooked in contact with an unsaturated polyester for 1.5 hours at 175 °C.

7.4 Aromatics from Photoinitiation Reactions and Photoinitiator Additives

Amongst the species considered here are: benzene, benzophenone, and the thio- or iodoaromatics from cationic photoinitiation, in addition to the photoinitiators themselves.

Benzene is a carcinogen and a potential by-product of cationic photoinitiator action. Benzene has been detected in packaging and foodstuffs [49] and studies have been carried out to determine benzene and other aromatics in headspace volatiles and toluene extracts of UV cured inks [50]. No benzene was detected, although a number of other aromatics compounds were detected, as shown below in **Table 16**.

Table 16. Aromatic compounds detected from UV cured ink			
Sample	By-product	Level (mg/m²)	Comments
Triarylsulfonium	Diphenyl sulfide	8.0	
Standard (Uvacure 1592)	Benzene	-	nd
Iodonium standard	Toluene	-	nd
IGM440	Iodotoluene	-	DNQ
Meerkat	Biphenyl	4.2	
	Isopropylthioxanthone	15.7	
Polecat	Biphenyl	-	DNQ
Bobcat	Isopropylthioxanthone (ITX)	9.1	
nd = Not detected by the method			
DNQ = Detected but not quantified			

With regard to the three compounds that have been detected at the highest levels, diphenyl disulfide (also called phenyl sulfide) is a severe irritant to the respiratory and digestive tract, and biphenyl is an irritant to both of these, but there is little information to hand on the acute or chronic toxicity of isopropylthioxanthone (ITX). Work that was reported on ITX as a result of it being detected in infant formula (see below) concluded that the existing *in vivo* genotoxicity studies do not indicate a genotoxic potential for ITX. The other initiator that was involved in this incident, 2 ethylhexyl-4-dimethylaminobenzoate (EHDAB), is not regarded a genotoxic or a teratogen.

The extraction medium is important for aromatic compounds of this type. For UV-cured epoxy films initiated with a mixed triarylsulfonium phosphate (Cyracure UVI-6990), much higher levels of extractables were observed with 50% ethanol rather than 10% ethanol or water: up to 272 ppb diphenyl sulfide was taken up in 50% ethanol [51]. HPLC with UV detection was used for these extract analyses. A method for the detection by GC of the photoinitiator residues in aqueous media has also been developed [35]. In this case, solid-phase extraction (on C_{18}-modified siloxanes) was used, followed by methanol desorption for the work up of the extracts.

Benzophenone is a photosensitiser, which may cause digestive tract irritation. GC-MS has been used to test for migration of this compound from cartonboard packaging [52]. In this study, 143 out of 350 samples of printed cardboard used with foodstuffs surveyed contained detectable benzophenone. In food, the highest level (7.3 mg/kg) was found for a high fat chocolate product packaged in direct contact, with 24% of the food samples surveyed having levels between 0.5 and 5.0 mg/kg. In those cases where there was no direct contact with the food, a six-fold reduction in the amount of benzophenone present was found.

Recently there have been reports of the photoinitiators, ITX and EHDAB, being found in products that are packed in printed cartons, such as milk and fruit juices. A notification from the Italian authorities under Article 50 of the Regulation (EC) No 178/2002 on the *Rapid Alert System for Food and Feed* showed the occurrence of ITX in liquid milk for babies at a level of 250 ug/l. As part of the investigation that ensued, the presence of EHDAB, often used as a synergist with ITX, was also revealed in food samples. Following a request from the Commission, a survey of ITX and EHDAB in a wide range of carton packed drinks, varying from water and fruit juices to flavoured milk, was carried out by industry. In this study, the full results of which are reported in *The EFSA Journal* [53] ITX was found to vary from 120 to 305 ug/l in milk products intended to be consumed by babies and very young children. No data on EHDAB was reported for these samples. In milk- and soya-based products the level of ITX ranged from 54 to 219 ug/l, and the level of EHDAB from 27 to 134 ug/l. Investigations at the time are believed to have revealed that the initiators found their way into the food products as a result of being transferred to the food contact site of the packaging by the phenomenon known as off-set (see Section 4.5). Details of EFSA's opinions on ITX and EHDAB can be found on their web site (*www.efsa.europa.eu*).

7.5 BPA and BADGE and Derivatives

BPA and, to a lesser extent, BADGE are of concern for endocrine disruption, and much effort has been devoted to the investigation of the possible migration of these from cured epoxies. BADGE is an important epoxy monomer and BPA is its immediate precursor. Whilst the focus of this work has been on direct food-contact coatings (i.e., can coatings), epoxies have widespread application in packaging as adhesives, coatings and binder resins for inks. Of course, the expectation is that direct food contact provides the worst-case scenario, where µg/kg levels of BPA and mg/kg levels of BADGE have been detected in certain foodstuffs [54-56].

Studies on the resin itself showed that BADGE migration levels decreased with increasing time of cure [57], whilst a further heating step reversed the trend [58]. The influence in thermal history seen here reflects the transition from kinetic to diffusion control in the progress of cure [59, 60], and impact of state of cure on migration behaviour [61].

Recently efforts have been made to improve analytical capability with respect to BADGE derivatives and related compounds. The former include low molecular weight BADGE oligomers, and the latter includes BFDGE.

HPLC coupled with UV, fluorescence and electrospray ionisation MS have all been used for the detection of BADGE oligomers of up to 1000 daltons [62]. The technique was applied to can coatings where a cumulative total of up 0.7 µg/dm^2 of coating was determined for BADGE-related species below 1000 daltons by acetonitrile extraction.

HPLC coupled with fluorescence detection has been used for the simultaneous determination of BADGE and BFDGE by two separate research groups [63, 64]. A limit of detection in various food simulants approaching ppb (µg/l) levels was claimed.

Excluding the oligomers from this analysis allows GC-MS to be used, and Czech researchers have recently reported the determination of BPA, BADGE, BPF, BFDGE by this technique [65]. The limits of detection in acetonitrile or food-stimulant extracts of coatings were below 1 µg/dm^2 for all four analytes. BPF and BFDGE were virtually undetectable in the coatings studied.

7.6 Epichlorohydrin

Epichlorohydrin (1-chloro-2,3-epoxypropane), is a carcinogen and a precursor to epoxy resin monomers. A method for the detection of epichlorohydrin in epoxy coatings by *n*-pentane extraction and analysis by GC with flame ionisation detection or selective-ion mass spectroscopy has been developed [66]. The respective limits of detection were 0.05 µg/ml for the former, and 0.02 µg/ml for the latter. However, no epichlorohydrin was detected.

The analysis of foodstuffs for epichlorohydrin has figured in one study where headspace GC-MS with selective ion detection was used to give a sensitivity of 0.02 mg/kg [67]. No epichlorohydrin was detected in this survey.

7.7 Bisphenol A

Although there is conflicting evidence regarding the safety of BPA, it has been targeted by organisations such as Friends of the Earth (FoE) and placed within a group of chemicals that are regarded as known, or suspected endocrine disruptors, and/or are bioaccumulative. FoE have been successful in persuading a number of major retailers (e.g., Marks and Spencer, Boots and B&Q) to sign a pledge committing them to identify products that contain such chemicals and phase them out by 2008 [68]. To help the retailers, FoE provided official lists, such as those issued by the Swedish, Danish and Dutch governments, and the companies used these to help themselves draw up a list of 15-20 priority substances. For example, one of the substances on the list were epoxy resin lacquers, containing BPA, which were used on the food contact side of the metal lids on food jars. These were phased out during 2004 and replaced by a non-epoxy resin lacquer [69].

7.8 Solvents

Where solvents are used, they are required for application purposes, and efficient drying would be expected to drive much of the solvent from the product. Nevertheless, migration of solvent residues from printing inks can be a possible source of off-flavour in food [70].

Given the wide volatility range of solvents exploited in the control of film formation, it might be expected that some of the higher boiling solvents (e.g., glycol ethers) might remain in the dried film product. Residual alkylbenzenes (C_{10}-C_{13} chain length) have been found to migrate into hamburger rolls and into a Tenax food simulant [71]. Printed hamburger collars had alkylbenzene contents in the range 70-500 mg/kg, and migration from the collar resulted in levels of 2 mg/kg in the hamburger rolls.

7.9 Plasticisers

Work by the Norwegian Food Safety Authority showed that a plasticiser used in ink formulations, *n*-ethyl-*o/p*-toluene-sulfonamide (N-ETSA), migrated into packaged cheese at a level of 13 mg/kg. Although there is no specific migration limit for N-ETSA in particular, European legislation prescribes a limit of 0.1 mg/kg for sulfonamides having a similar chemical structure. The level found by the Norwegians is obviously significantly higher than this, but these types of problems can be avoided by the use of polymeric type plasticisers/flexibilisers that have a greatly reduced potential to migrate. This targeted piece of research is complemented by a surveillance exercise of food packaging materials undertaken by the same authority. When a printed laminate was tested, N-ETSA was found to have migrated into the water and oil olive simulants. Other plasticiser and ink related species were also detected and quantified [72].

7.10 Extractables from UV-Cured Coating for Cardboard

Gaube and Ohlemacher [73] have reported on the parameters that affect the extractables of cationic UV-cured coatings that are applied to cardboard. Specific migration and extraction experiments were carried out to determine the affect of changing parameters such as pre-treatment of the cardboard, and the formulation of the coating, on the concentration of the cationic photoinitiator (bis[4-(diphenylsulfonio)-phenyl]sulfide-bis-hexafluorophosphate) and the epoxy monomer (3,4-epoxycyclohexylmethyl-3′,4′-epoxycyclohexane carboxylate).

7.11 Potential Migrants

A recent research project looking at the potential migrants present in coatings and inks used on the non-food contact side of food packaging was carried out at Rapra Technology [17]. This project was sponsored by the UK FSA and, as a part of the final report, important species that had the potential to migrate were tabulated. This information is reproduced in **Table 17**. A subjective importance rating based on toxicity, potential abundance, mobility and molecular weight is provided in **Table 17**.

In many coatings and inks products, the drying (curing) mechanisms rely on reactive systems and the reaction products are included in **Table 17**. Some substances such as styrene, BADGE and the photoinitiator ITX have already been the subjects of migration studies (see Section 7.4). Other important by-products of initiator action include benzaldehyde, benzene, biphenyl and diphenyl sulfide.

Aromatic amines are used in the manufacture of azo dyes and pigments, and amines are also potential breakdown products from the hydrolysis of amides, unreacted isocyanates or the action of heat on cationic surfactants, all used with inks and coatings. Commission Directive 2002/72/EC details that materials and articles manufactured by using aromatic isocyanates or colorants prepared by diazo-coupling, shall not release primary aromatic amines (expressed as aniline) in a detectable quantity - detection limit = 0.02 mg/kg of food or food simulant, analytical tolerance included. In the list of possible important impurities and breakdown products in **Table 17**, fifteen non-permitted aromatic amines are detailed. A further important precursor/pigment breakdown product that is included is the potentially genotoxic substance nitrotoluene.

Many starting substances currently used in food use coatings and inks are not listed in Inventory lists (e.g., such as the Council of Europe's – see Section 5.2) and have not yet been evaluated by the European Food Safety Authority (EFSA).

Table 17. Possible impurities and breakdown products identified by the FSA Coatings and Inks Project

Listed in this Table are species not included in either the CoE or BCF Inventory lists. The entries here are based on the chemistry discussed in the final FSA report (copy available from the FSA – see Section 10), and reference should be made to the appropriate section of this report for due context. This compilation is intended to illustrate some of the possibilities and is not intended to be exhaustive. A subjective importance rating based on toxicity, potential abundance, mobility, and molecular weight are also provided. (1 = High, 2 = Medium, 3 = Low)

CAS No.	Substance	ADI/TDI or SML etc.	Source	Relative importance
79-06-1	Acrylamide	Not established. Possible carcinogen	Precursor to polyacrylamide (water-soluble binder)	1
95-23-8	5-Aminobenzimidazolone, (5-amino-1,3-dihydro-2H-benzimidazol-2-one)	Aromatic amine. SML = Not detectable	Precursor/breakdown product of PY 194	1
873-74-5	4-Aminobenzonitrile	Aromatic amine. SML = Not detectable	Possible breakdown product of PY 181	1
62-53-3	Aniline	Aromatic amine. SML = Not detectable	Precursor/breakdown product of PR 2, PY 1 & PY 12	1
71-43-2	Benzene	Carcinogen	Possible breakdown product of sulfonium photoinitiator. Possible impurity in unsaturated polyester binders	1
25834-80-4	2,4-Bis(*p*-amino-benzyl)aniline	Aromatic amine. SML = Not detectable	Possible precursor/breakdown product of PU binders	1

Table 17. Continued

CAS No.	Substance	ADI/TDI or SML etc.	Source	Relative importance
–	Bis(thiophenyl)benzene	Lack of information	Possible breakdown product of sulfonium photoinitiator	1
6358-64-1	4-Chloro-2,5-dimethylaniline	Aromatic amine SML = Not detectable	Precursor/breakdown product of PY 83	1
101-77-9	4,4'-Diaminodiphenyl-methane, (4,4 methylene-dianiline)	Aromatic amine SML = Not detectable	Possible precursor/breakdown product of PU binders. Possible curing agent in epoxy binders	1
91-04-1	3,3'-Dichlorobenzidine	Aromatic amine SML = Not detectable	Precursor to PY 12, 13, 14, 17, 55 & 83	1
119-90-4	3,3'-Dimethoxybenzidine, (o-dianisidine)	Aromatic amine SML = Not detectable	Precursor to PO 16	1
95-68-1	2,4-Dimethylaniline, (2,4-xylidine)	Aromatic amine SML = Not detectable	Precursor/breakdown product of PY 13	1
94-70-2	2-Ethoxyaniline, o-phenetidine	Aromatic amine SML = Not detectable	Precursor/breakdown product of PR 170	1
90-94-0	2-Methoxyaniline, o-anisidine	Aromatic amine SML = Not detectable	Precursor/breakdown product of PY 17	1
95-53-4	2-Methylaniline (o-toluidine)	Aromatic amine SML = Not detectable	Precursor/breakdown product of PR 12, PY 14	1

Table 17. *Continued*

CAS No.	Substance	ADI/TDI or SML etc.	Source	Relative importance
106-49-0	4-Methylaniline (*p*-toluidine)	Aromatic amine SML = Not detectable	Precursor/breakdown product of PY 55	1
99-08-1	*m*-Nitrotoluene (3-nitrotoluene)	Possibly some genotoxicity	Possible breakdown product of PY 1, PR 3 & PR 12	1
95-80-7	2,4-Toluenediamine, (2,4 diaminotoluene)	Aromatic amine SML = Not detectable	Possible precursor/breakdown product of PU binders	1
823-40-5	2,6-Toluenediamine, (2,6 diaminotoluene)	Aromatic amine SML = Not detectable	Possible precursor/breakdown product of PU binders	1
2835-68-9	*p*-Aminobenzamide	SML = 0.05 mg/kg (only to be used in PET)	Precursor/breakdown product of PY 181	2
100-66-3	Anisole (methoxybenzene), (methyl phenyl ether)	Lack of information	Possible breakdown product of PY 194	2
55-21-0	Benzamide	Not in 2002/72/EC	Possible breakdown product of PR 170	2
92-52-4	Biphenyl	Fungicide used with oranges	Possible breakdown product of sulfonium photoinitiator	2
2409-55-4	2-*Tert*-butyl-4-methylphenol	Relatively common breakdown product	Possible breakdown product of BHT (antioxidant)	2
2460-77-7	2,5-Di-*tert*-butyl-1,4-benzoquinone	Lack of information	Possible breakdown product of BHT	2

Table 17. *Continued*

CAS No.	Substance	ADI/TDI or SML etc.	Source	Relative importance
1620-98-0	3,5-Di-*tert*-butyl-4-hydroxybenzaldehyde	Lack of information	Possible breakdown product of BHT	2
106-46-7	*p*-Dichlorobenzene, (1,4-dichlorobenzene)	SML = 12 mg/kg (2002/72/EC)	Possible breakdown product of PR 2	2
–	2,4-Diethyl-9H-thioxanthen-9-ol	Lack of information	Possible breakdown product of DETX	2
91-01-0	Diphenyl-2-methanol (benzhydrol)	Limited information found	Possible breakdown product of benzophenone	2
139-66-2	Diphenyl sulfide (phenyl sulfide)	Lack of information	Possible breakdown product of sulphonium photoinitiator	2
4088-22-6	Distearylmethylamine	Lack of information	Possible breakdown product of dimethyldioctadecyl-ammonium chloride	2
100-41-4	Ethylbenzene	Common impurity in styrenics	Possible impurity in unsaturated polyester binders	2
–	4,4'-Ethylenebis(2,6-di-*tert*-butylphenol)	Lack of information	Possible breakdown product of BHT	2
–	4-Hydroxymethylphenyl-4'-methyldiphenyl sulfide	Lack of information	Possible breakdown product of 4-benzoyl-4'-methyldiphenyl sulfide	2

Table 17. *Continued*

CAS No.	Substance	ADI/TDI or SML etc.	Source	Relative importance
92-70-6	β-Hydroxynaphthoic acid, (3-hydroxy-2-naphthoic acid), (2-Hydroxy-3-naphthoic acid)	Lack of information	Precursor/breakdown product of PR 48:2	2
–	p-Iodoisobutylbenzene	Lack of information	Possible breakdown product of diaryliodinium photoinitiator	2
624-31-7	p-Iodotoluene	Lack of information	Possible breakdown product of diaryliodinium photoinitiator	2
538-93-2	Isobutylbenzene (2-methyl-1-phenylpropane)	Lack of information	Possible breakdown product of diaryliodinium photoinitiator	2
2855-13-2	Isophorondiamine (5-amino-1,3,3-trimethyl-cyclohexane-methylamine)	Lack of information	Possible precursor/breakdown product of PU binders	2
–	2-Isopropyl-9H-thioxanthen-9-ol	Probably considered in EFSA review of ITX	Possible breakdown product of ITX	2
124-09-4	Hexamethylenediamine (1,6-hexanediamine)	SML = 2.4 mg/kg	Possible precursor/breakdown product of PU binders	2
26266-77-3	Hydroabietyl alcohol	SCF list 8	Possible modifier in alkyd binders	2

Table 17. *Continued*

CAS No.	Substance	ADI/TDI or SML etc.	Source	Relative importance
–	5-Methylbenzimidazolone (5-methyl-1,3-dihydro-2H-benziimidazol-2-one)	Lack of information	Possible breakdown product of PO 64	2
1761-71-3	4,4'-Methylenebis(cyclo-hexylamine)	Lack of information	Possible precursor/breakdown product of PU binders	2
135-19-3	β-Naphthol (2-hydroxynaphthalene)	Lack of information	Precursor/breakdown product of PR 3	2
126-30-7	Neopentyl glycol (2,2-dimethyl-1,3-propanediol)	SML = 0.05 mg/kg in 2002/72/EC	Possible precursor to polyester binder. Precursor/breakdown product of neopentyl glycol plasticisers	2
104-40-5	4-Nonylphenol	TDI (NP) = 0.005 mg/kg body weight/day	Precursor to ethoxylated nonylphenol surfactants	2
11066-49-2	Isononylphenol	TDI (NP) = 0.005 mg/kg body weight/day	Precursor to ethoxylated nonylphenol surfactants	2
140-66-9	4-(*tert*-octyl)phenol, [4-(1,1,3,3-tetramethyl-butyl) phenol]	Lack of information	Precursor to ethoxylated octylphenol surfactants	2
85-41-6	Phthalimide	Lack of information	Precursor/breakdown product of PB 15 and 15:1-15:6	2

Table 17. *Continued*

CAS No.	Substance	ADI/TDI or SML etc.	Source	Relative importance
638-65-3	Stearonitrile	Lack of information	Possible breakdown product of stearic acid amide	2
–	3,3′,5,5′-tetrabis(*tert*-butyl)-4,4′-stilbenequinone	Lack of information	Possible breakdown product of BHT	2
108-88-3	Toluene	Some toxicity	Possible breakdown product of diaryliodinium photoinitiator. Possible impurity in unsaturated polyester binders	2
124-04-9	Adipic acid	Listed in 2002/72/EC. No SML	Possible precursor to alkyd binders	3
100-52-7	Benzaldehyde	Listed in 2002/72/EC. Warning given re: possible tainting of food	Breakdown product of DMPA initiator. Possible impurity in unsaturated polyester binders	3
91-76-9	Benzoguanamine (2,4-Diamino-6-phenyl-1,3,5-triazine)	QMA = 5 mg/6 dm^2	Possible precursor to amino binders	3
74-87-3	Chloromethane (methyl chloride)	bp – 24 °C	Possible breakdown product of cationic surfactants	3
461-58-5	Cyanoguanidine (dicyanodiamide)	Listed in 2002/72/EC. No SML	Possible curing agent in epoxy binders	3

Table 17. *Continued*

CAS No.	Substance	ADI/TDI or SML etc.	Source	Relative importance
–	*N,N*-Dimethylcoco-alkylamine	Not listed in 2002/72/EC but probably of low toxicity	Possible breakdown product of (cocoalkyl)trimethyl-ammonium chloride	3
–	Di-*n*-octyltin mono(2-ethylhexyl mercaptoacetate) monochloride	Controlled by SML(T) of 0.006 mg/kg expressed as tin in 2005/79/EC	Possible breakdown product of di-*n*-octyltin bis(2-ethylhexyl mercaptoacetate (in PVC)	3
693-23-2	Dodecandioic acid, (1,10-decanedicarboxylic acid)	In 2002/72/EC No SML	Possible precursor to polyester binder	3
104-76-7	2-Ethyl-1-hexanol (isooctyl alcohol)	Low toxicity	Precursor to 2-ethylhexyl phthalate plasticisers	3
110-17-8	Fumaric acid	Listed in 2002/72/EC. No SML	Possible precursor to alkyd binders	3
97-65-4	Itaconic acid (methylenesuccinic acid)	Listed in 2002/72/EC. No SML	Possible impurity in water-soluble binders	3
108-78-1	Melamine	SML = 30 mg/kg in 2002/72/EC	Possible precursor to amino binders	3
–	Mono-*n*-octyltin bis(2-ethylhexyl mercaptoacetate) monochloride	Controlled by SML(T) of 0.006 mg/kg expressed as tin in 2005/79/EC	Possible breakdown product of mono-*n*-octyltin tris(2-ethylhexyl mercaptoacetate) (in PVC)	3

Table 17. Continued

CAS No.	Substance	ADI/TDI or SML etc.	Source	Relative importance
–	Mono-*n*-octyltin mono(2-ethylhexyl mercaptoacetate) dichloride	Controlled by SML(T) of 0.006 mg/kg expressed as tin in 2005/79/EC	Possible breakdown product of mono-n-octyltin tris(2-ethylhexyl mercaptoacetate) (in PVC)	3
–	Oxides of nitrogen	Nitrous oxide is used as a propellant gas with food products	Possible breakdown product of nitrocellulose binders	3
111-20-6	Sebacic acid	Listed in 2002/72/EC. No SML	Possible precursor to alkyd binders	3
50-70-4	Sorbitol	Listed in 2002/72/EC. No SML	Possible precursor to alkyd resins	3
77-99-6	Trimethylolpropane	SML = 6 mg/kg in 2002/72/EC	Possible precursor to alkyd resins	3

ADI: *Acceptable daily intake*
TDI: *Tolerable daily intake*
DETX: *2,4-diethylthioxanthane*
BHT: *Butylated hydroxyl toluene*
NP: *Nonyl Phenol*
QMA: *Maximum permitted quantity of substance in the finished product*
bp: *Boiling point*
SML(T): *Total of substances/moieties listed*
DMPA: *2,2 Dimethoxy-2-phenylacetophenane*
PVC: *Polyvinylchloride*

8 Improving the Safety of Inks and Coatings for Food Use

8.1 New Food Approved Pigments

Work is on-going to widen the range of pigments that can be used for food contact applications, and a steady stream of new products from the manufacturers is being brought onto the market. In addition to providing alternatives to pigments that do not have food approval, other improvements cited are the ability to used in higher loadings and the wider range of manufacturing processes and conditions for which the pigments are compatible [76].

8.2 Water-Based Systems

The exposure of humans to phthalate ester compounds has been a concern now for many years, with the first major press release by MAFF warning of the potential for migration from packaging materials, particularly cling film, coming in the middle of the 1980s. The most widely used phthalate, di-ethylhexyl phthalate has been linked in animal studies to damage to the kidneys and liver, and has been labelled as a probable human carcinogen [77, 78]. Phthalate plasticisers are still used in some coating and ink formulations and this has assisted the development of a new generation of water-based coatings, which are free of phthalates [79]. In addition, these coatings, which can be used for a wide range of applications, have the additional benefit of being free from VOC, which assists manufacturers in their endeavours to meet environmental emission targets for these types of compounds. However, there is now a further move, to newer technology using UV/EB curing systems (see Section 8.3).

8.3 UV/EB Curable Systems

For a number of years now there has been a move away from water-and solvent-based coating and ink systems and towards the use of formulations that can be cured by utilising either the energy produced by an UV source or an EB. In addition to offering advantages in ease of handling, superior flow characteristics and low odour, the technology results in a lower level of potential migrants in the final product – an obvious benefit for food contact applications [80, 81]. However, although offering some common benefits, the two technologies are quite different and the EB curing mechanism is regarded as having a number of advantages over the ultraviolet light curing mechanism for food packaging applications [82, 83].

Examples of these advantages include: a higher degree of reaction (giving a lower level of extractables), a higher processing speed, and that fact that no initiator breakdown products are generated.

Despite the advantages that these techniques offer, there are still some manufacturing problems that are encountered. For example, a problem that can be experienced when off set printers use UV ink technology is poor adhesion to the substrate (either plastic or metal), due to either poor wetting or film shrinkage, or a combination of the two. This is being addressed by applying modifications to the technology (e.g., the development of new resins), and by raising the surface energy of certain substrates (e.g., polyolefins) by using in-line techniques such as Corona discharge treatments [84].

To facilitate the expansion and acceptance of UV/EB technology for food contact products, the RadTech Food Packaging Alliance has been formed and it has sponsored the production of migration data on a number of substances used in UV/EB curable coatings, ink and adhesive materials [85].

8.4 New Initiators for UV Curable Inks

As mentioned previously, UV curable inks are gaining in popularity over conventional solvent and water based inks. Although this technology offers a number of benefits, it is appreciated that one area where there is scope for more work is in the development of a greater range of initiators that do not produce breakdown and reaction products that can cause taint and odour problems in food contact applications [86].

9 Future Trends

In addition to the constant improvements in the safety of coatings and inks, which are described in Section 8, there are also a number of technological improvements surrounding the coatings and inks industry that are worth mentioning. Some principal examples of these, together with the citation of some relevant recently published literature, are covered next.

9.1 Improvements in Recycling Systems

Although the use of multi-layer, laminate products for the packaging of food can cause problems when it comes to recycling them at the end of their life (see Section 9.2), systems are being developed that can cope with packaging that has a coating or is multi-layered. An example of this is a plant that can recycle polyethylene terephthalate (PET) beer bottles to generate a food grade resin that can be re-used in drinks bottles [87].

9.2 Biodegradability

Biodegradability and compostibility are becoming increasing attractive attributes for all food contact products, as the need to reduce the amount of waste that is placed into conventional landfill sites. This is a particularly advantageous property for products that are made up of a number of different polymer based components, which makes recycling an unattractive option due to separation problems.

An example of the new, multi-component biodegradable products that are coming onto the market is lunch box sheet, where the film, adhesive and ink are all biodegradable. This enables the complete product to be placed into a composter for disposal [88].

9.3 Use of Coatings to Improve Barrier Properties of Food Packaging

Coatings can be used to improve the barrier properties of food packaging films, and hence increase the shelf life of food and beverages. In addition to decreasing the amount of permeation that takes place, these coatings can also have secondary benefits when used in laminating products, such as improving the interlayer adhesion [89].

Nanotechnology is being incorporated into a wide range of manufacturing sectors, and one application of it in food packaging is to improve the barrier properties of films. This can be achieved by the use of thin polymer films that contain nanoclay particles [90].

Grande has recently discussed the latest developments in barrier technologies, including coatings, with respect to polyethylene terephthalate polyester bottles [91].

9.4 Antimicrobial Systems

Interest in the use of antimicrobial products for the food industry is growing. Their has been a lot of activity in the development of antimicrobial additives for food contact rubbers and, in order to create a antimicrobial coating for rubbers, silver nanoparticles have been deposited onto the surface of food contact silicone rubbers [18]. These types of products are also being developed for the coatings and inks industries. A particular example is a white pigmented powder coating type paint for the food processing industry that has antimicrobial functionality due to the presence of silver ions [92]. The use of a silver glass ceramic antimicrobial additive for the inner surfaces of refrigerators has already been mentioned in Section 3.3.

In the food packaging area, antimicrobial technologies (e.g., in the form of an internal coating) have the potential to extend the shelf life of perishable products. In the case of the food processing industry, paints are being developed that contain anti-microbial additives such as silver glass and silver zeolite [92].

The anti-microbial properties of paper and board type food packaging materials can be improved by the use of polymer binder solutions that have been treated with Nisin, a bacteriocin produced by *Lactococcus lactis*. This natural anti-microbial agent has the benefits of being able to withstand both high temperature processing and acidic environments and, being a hydrophobic protein, has 'Generally recognised as safe' status in the USA for use in cheese products [93]. In addition to Nisin, chitosan (a polysaccharide and deacetylated form of chitin) also shows promise as a natural antimicrobial for use in paper binders [94].

Gergely has recently reviewed the regulatory situation in the EU with respect to additives in food contact materials that exhibit antimicrobial activity [95].

9.5 Laser Marking to Replace Conventional Inks

Continual improvements in laser technology have resulted in a system for plastics that offers indelibility coupled with high speed [96]. Given that a laser pigment for this type system has received FDA approval, this type of technology could begin to mount a credible challenge to conventional inks in the future.

9.6 Intelligent and Active Packaging

A definition of intelligent packaging is the kind of packaging that uses devices within the pack or as part of the package itself to sense and register certain changes in the pack and its contents. Areas of intelligent packaging that are attracting an increasing amount of attention are the development of time-temperature indicators and, for foods that have been packaged under inert environments, oxygen sensors.

In the case of oxygen sensors, it is important that they are non-evasive and an example of the type of product which is of relevance in this report is the use of an oxygen sensor that can be printed as an ink onto packaging [97].

Active packaging is the term used to describe packaging products that are not simply passive protectors of the food within them, but interact with the product to maintain its integrity and increase its shelf life. Examples of active packing include [98]:

a) The scavenging of ethylene to slow the ripening of fruits and vegetables

b) The scavenging of oxygen to prevent things such as the development of odours, changes in colour, and mould growth.

c) Use of gases such as carbon dioxide or sulfur dioxide to prevent microbiological growth.

9.7 Applications of Nanotechnology

The use of nanotechnology to improve the barrier properties of food packaging has been mentioned in Section 9.3. Another area where nanotechnology shows considerable promise is where nanoparticles in inks can be used to improve the capability of radio frequency identification technology. In addition to conventional bar code type information, this would also enable addition information such as the state of the food in the packaging to be accessed. Another area that is under development and evaluation is the use of nanoparticles as pigments in inks. The use of nanotechnology to produce intelligent packaging is also being investigated by a team at Strathclyde University [98].

It was the increasing use, and potential use in the future, of nanotechnology in food packaging applications that led the FSA to fund a project to assess the impact that this technology could have on the safety of food contact materials. Information on this project, which started in 2005, can be found on the FSA's web site (see Sources of Information.

9.8 Developments in Analytical Techniques

Analytical chemistry plays a vital role in the assessment of food safety of all food contact materials and is invaluable in the determination of the specific migration behaviour of selected, targeted species. For many years, HPLC was, in practice, the only available technique for the determination of thermally labile and/or relatively high molecular weight migrant species. However, in the past five years or so LC-MS instruments have proliferated to the extent that they have now replaced HPLC in the majority of laboratories. These instruments are a much better complement for GC-MS, than HPLC, and enable the analyst for the first time to routinely generate data on a full range of compounds (i.e., thermally labile, stable and polar substances, such as salts) up to 1000 daltons, which is the established upper limit for chemical absorption in the gastrointestinal tract. LC-MS enables more accurate conformity checks to be performed on coatings and inks formulations, as well as adding to the understanding of the migration behaviour of their low molecular weight constituents and reaction and breakdown products.

Development work also continues to provide analytical instrumentation which offers commercially accessible improvements in important parameters such as molecular weight range, detection limits, software assisted peak deconvolution, analysis speed, accuracy of library searching and species selectivity. The introduction of mid-priced multi-hyphenated techniques such as GCxGC-time-of flight MS and LC-MSxMS are examples of this. In addition to contributing to conformity work, and the analysis of food simulants, these instruments with their greater resolving power and selectivity are also improving routine, direct analysis of food products, where the potentially large range of low molecular weight species can cause interference problems.

10 Conclusion

This review has provided the reader with an overview of the types of coatings and inks that are used for food contact materials, an introduction to the technology that is associated with their manufacture and an overview of the legislation that is associated with these types of products. It has also provided a summary of the migration data that is available and a description of the advances that are being taken by industry to improve the safety and functionality of these types of products.

Food contact legislation for coatings and inks within the EU has been a very active area recently, with both a CoE resolution on coatings, and one for inks for non-direct food use, having been adopted within the last couple of years. The latter document has proved to be controversial, with industry groups such as the BCF and EuPIA being far from happy with it, and providing their own inventory list and guideline documents in response.

For the analysis of coatings and inks, the commercial proliferation of LC-MS instruments, with their enhanced capability compared to the much older HPLC technique, and complimentary status to GC-MS, will be of great benefit to analysts who are carrying out tasks ranging from reverse engineering and failure diagnosis to migration studies. This technique will enable all of this work to be carried out with greater ease and effectiveness, and may assist in the work required to generate EU regulations for both coatings and inks.

The product groups within the inks and coatings sector are experiencing a number of important developments. Some of these address environmental concerns, such as the move from solvent-based systems to water-based products, and the increased use of biodegradable materials, whereas others, such as the use of nano-materials, the use of antimicrobial agents, and those to produce active/intelligent packaging, are designed to create products that provide the customer and the retailer with products having greater degrees of safety and quality.

Coatings and inks for food contact materials will continue to be a very active and dynamic area of the polymer industry for the foreseeable future.

Sources of Further Information and Advice

There are a number of routes that a researcher can take to obtain further information. It is not possible within this format to provide a comprehensive list, but this section provides a summary of the key areas where knowledge can be found, with a number of examples included in each category.

Reference Books

1.   T.A Turner, *Canmaking: The Technology of Metal Protection and Decoration, Crown Cork and Seal*, Blackie Academic and Professional, London, UK, 1997.

2.   *Manual for Resins for Surface Coatings, Volume 2*, Eds., P. Oldring and G. Haywar, SITA Technology, London, UK, 1987.

3.   T. Hutton, *Food Manufacturing: An Overview*, CCFRA, Chipping Camden, UK, 2001.

4.   T. Hutton, *Introduction to Food Hygiene in Food Processing*, CCFRA, Chipping Camden, UK, 2007.

5.   T.Hutton, *Food Packaging*, CCFRA, Chipping Camden, UK, 2003.

6.   N. Anyadike, *Introduction to Flexible Packaging*, PIRA, Leatherhead, UK, 2003.

7.   *The Printing Ink Manual, 5th Edition*, Eds., R.H Leach and R.J Pierce, Kluwer Academic Publishers, Dordrecht, The Netherlands, 2002.

8.   P. Oittinen and H. Saarelma, *Printing*, Fapet Oy, Helsinka, Finland, 1998.

9.   *Lacquers, Varnishes and Coatings for Food and Drink Cans and for the Metal Decorating Industry*, ICI Packaging Coatings, 2000.

10.   *Chemical Migration and Food Contact Materials*, Eds., K.A. Barnes, C.R. Sinclair and D.H. Watson, Woodhead Publishing, Cambridge, UK, 2006.

11.   *Migration from Food Contact Materials*, Ed., L.L. Katan, Blackie Academic and Professional, London, UK, 1996.

12.   *Additives for Coatings*, Ed., J. Bieleman, Wiley-VCH, Weinheim, Germany, 2000.

13.   *Chemistry and Technology of UV and EB Formulation for Coatings, Inks and Paints, Volume II: Prepolymers and Reactive Diluents*, Ed., G. Webster, John Wiley, Chichester, UK, 1997.

14.   J.P. Dowling, P. Pappas, B. Monroe and A. Carroy, *Chemistry and Technology of UV and EB Formulation for Coatings, Inks & Paints, Volume V: Speciality Finishes*, Wiley, Chichester, UK, 1997.

15.   B. Thomson, *Printing Materials: Science and Technology*, PIRA International, Leatherhead, UK, 2004.

Reports

1.   Food Standard Agency Reports

Food Standard Agency/MAFF Projects on Food Contact Rubbers and Rubber Latex:

A03043 – J. Haines and co-workers, Assessment and quantification of latex protein (LP) transfer from

LP-containing materials into food and drink products, 2004.

A03038 – M.J. Forrest and co-workers, Rubber Breakdown Products, 2005.

FS2248 – J.A. Sidwell and co-workers, Further Migration Data on Food Contact Rubbers, 1997.

FS2219 – J.A. Sidwell and co-workers, Migration Data on Food Contact Rubbers, 1994.

M.J Forrest and co-workers, Food Standards Agency Project A03046 – *Chemical migration from silicones used in connection with food contact materials and articles*, 2005.

E. Bradley, Combined Food Standards Agency Projects FS2251 and A03022 – Overall Title: *A systematic investigation into potential chemical migration from inks and associated coatings used on the food contact surface of packages*, 2002.

2. British Coatings Federation, *The BCF guide to printing inks for use on food wrappers and packages*, Leatherhead, UK, 2002.

3. C. Brede, I. Skjevrak and P. Fjeldal, *SNT Arbeidsrapport*, 2003, 3.

Professional, Research, Trade and Governmental Organisations

Council of Europe – Partial Agreement in the Social and Public Health Field, *www.coe.int/soc-sp*

UK Food Standards Agency (FSA), *www.foodstandards.gov.uk*

US Food and Drug Agency, *www.fda.gov*

Bundesinstitut fur Risikobewertung (BfR) (German Federal Institute for Risk Assessment), *www.bfr.bund.de*

The European Council of Paint, Printing Inks and Artists' Colours Industry (CEPE), *www.cepe.org*

European Printing Ink Manufacturers Association (EuPIA), *www.eupia.org*

European Food Safety Association (EFSA), *www.efsa.europa.eu*

Institute of Materials, Minerals and Mining (IOM[3]), *www.iom3.org*

Leatherhead Food International, *www.leatherheadfood.com*

Central Science Laboratory, *www.csl.gov.uk*

PIRA International, *www.pira.co.uk*

Fraunhofer Institut Angewandte Polymerforschung, *www.pioneers-in-polymers.com*

Commercial Abstract Databases

a) Rapra Abstracts (The Polymer Library) – Rapra Technology

b) PIRA Abstracts – Pira International

c) Chemical Abstracts – American Chemical Society

d) World Surface Coatings Database – PRA Coatings Technology Centre

Acknowledgements

The author would like to acknowledge the contribution made to this report by Dr Bryan Willoughby and Dr Ray Good. This arose out of the work that they were commissioned to do by Rapra Technology for the FSA Coatings and Ink project (Contract A03055). In particular, Dr Willoughby's work been used in Chapters 2, 4 and 7 and **Table 17**, and Dr Good's industry survey has been used as the basis for Chapter 3 and **Tables 5 to 14**.

References

1. B.G. Willoughby, *Air Monitoring in the Rubber and Plastics Industries*, Rapra Technology Ltd., Shawbury, Shrewsbury, UK, 2003.

2. D.J. Lyman, *Polyurethanes: The Chemistry of the Diisocyanate-Diol Reaction*, Dekker, Ed., D.H. Solomon, New York, NY, USA, 1972, p.95.

3. M.J. Forrest and co-workers, *Food Standards Agency Project A03046 – Chemical migration from silicones used in connection with food contact materials and articles*, Food Standards Agency, London, UK, 2003-2005.

4. D.L. Stalker and F. Sandmeyer, *Surface Coatings International Part B*, 2002, 85, B4, 309.

5. R. Good, *Contribution to Food Standards Agency Project A03046 – Chemical migration from silicones used in connection with food contact materials and articles*, Food Standards Agency, London, UK, 2003-2005.

6.   R. Good, *Presentation on Speciality Coatings and Varnishes for Metal & Flexible Packaging, Food Packaging Interactions*, Campden and Chorleywood Food Research Association, Chipping Campden, UK, 2005.

7.   K. Johns in *Proceedings of Silicone in Coatings IV*, Guildford, UK, 2002, p.13.

8.   S. Notermans and E. Hoornstra in *Proceedings of a Paint Research Association Conference on Hygenic Coatings*, Brussels, Belgium, 2002, p.8.

9.   *WHO Surveillance Programme for Control of Foodborne Infections and Intoxications in Europe*, 7th Report 1993-1998, Eds., C. Tirado and K. Schmidt, FAO/WHO Collaborating Centre for Research and Training in Food Hygiene and Zoonoses, Berlin, Germany, 2000.

10. Trade Literature, Stirling Lloyd Polychem Ltd, Knutsford, UK, 2002.

11. H. Studer in the *Proceedings of Addcon World 2005 - the 11th Rapra International Plastics Additives and Modifiers Conference*, Hamburg, Germany, 2005, p.13.

12. E.L. Bradley, L. Castle, T.J. Dines; A.G. Fitzgerald, P. Gonzalez Tunon, S.M. Jickells, S.M. Johns, E.S. Layfield, K.A. Mountfort, H. Onoh and I.A. Ramsay, *Food Additives and Contaminants*, 2005, **22**, 5, 490.

13. D.J. Knight and L.A. Creighton, *Regulation Of Food Packaging in Europe and the USA*, Rapra Review Report, Rapra Technology Ltd., Shawbury, Shrewsbury, UK, 2004, **173**, 15, 5.

14. J. Sidwell in the *Proceedings of the 12th International Rapra Plastics Additives and Modifiers Conference - Addcon World 2006*, Cologne, Germany, 2006, Rapra Technology, 2006, p.2.

15. C. Whitehead in *Proceedings of European Food Packaging Regulations: Support from your Printing and Coating Suppliers*, EuPIA seminar, Solihull, UK, 2006.

16. M.J. Forrest and co-workers, Food Standards Agency Project A03055 – *An Assessment of the Potential of Migration of Substances from Inks and their Associated Coatings*, Food Standards Agency, London, UK, 2005-2007.

17. M. Forrest in *Chemical Migration and Food Contact Materials*, Eds., K.A. Barnes; C.R. Sinclair and D.H. Watson, Woodhead Publishing, Cambridge, UK, 2007, 271.

18. R. Ashby, I. Cooper, S. Harvey and P. Tice, *Food Packaging Migration and Legislation*, PIRA International, Surrey, UK, 1997,

19. V. James, *Polymers Paint Colour Journal*, 2006, **196**, 4504, 48.

20. A. Lin, H. Gao, G. Wind and F. Wornick in *Proceedings of the Tappi 2002 PLACE Conference*, Boston, MA, USA, 2002, Session 13, 48.

21. D. Scholler, J-M. Vergnaud, J. Bouquant, H. Vergallen and A. Feigenbaum, *Packaging Technology and Science*, 2003, **16**, 5, 209.

22. A. Feigenbaum, D. Scholler, J. Bouquant, G. Brigot, D. Ferrier, R. Franz, L. Lillemark, A.M. Riquet, J.H. Petersen, B. van Lierop and N. Yagoubi, *Food Additives and Contaminants*, 2002, **19**, 2, 184.

23. K. Grob, *Food Additives and Contaminants*, 2002, **19**, Supplement, 185.

24. CEN/TS 13130-25, *Materials and Articles in Contact with Foodstuffs - Plastics Substances Subject to Limitation - Part 25: Determination of 4-Methyl-1-Pentene in Food Simulants*, 2005.

25. CEN/TS 13130-13, *Materials and Articles in Contact with Foodstuffs - Plastics Substances Subject to Limitation - Part 13: Determination of 2,2-Bis(4-Hydroxyphenyl)Propane (Bisphenol A) in Food Simulants*, 2005.

26. CEN/TS 13130-26, *Materials and Articles in Contact with Foodstuffs - Plastics Substances Subject to Limitation - Part 26: Determination of 1-Octene and Tetrahydrofuran in Food Simulants*, 2005.

27. EN 13130-11, 2004, Materials and Articles in Contact with Foodstuffs – *Plastics Substances Subject to Limitation - Part 1: Guide to Test Methods for Specific Migration*, 2004.

28 . R. Rijk and R. Bas in *Proceedings of a Pira International Conference - Plastics in Contact with Foodstuffs*, London, UK, 2000, p.8.

29. M. Forrest, S. Holding and D. Howells in the *Proceedings of a Rapra Technology Ltd., Conference - High Performance and Speciality Elastomers 2005*, Geneva, Switzerland, 2005, p.2.

30. R. Veraart and L. Coulier in *Chemical Migration and Food Contact Materials*, Eds., K.A. Barnes, C.R. Sinclair and D.H. Watson, Woodhead Publishing, Cambridge, UK, 2007, p.87.

31. S.M. Jickells, C. Crews, L. Castle and J. Gilbert, *Food Additives and Contaminants*, 1990, 7, 2, 197.

32. S. Paz-Abuin, A. Lopez-Quintela, M. Pazos, P. Prendes, M. Varela and P. Paseito in *Proceedings of the Epoxy Technologies for Ambient Cure Protective Coatings Conference*, Brussels, Belgium, 1997, p.23.

33. J. Salafranca and R. Franz, *Deutsche Lebensmittel Rundschau*, 2000, 96, 10, 355.

34. S. Papilloud and D. Baudraz, *Food Additives and Contaminants*, 2002, 19, 2, 168.

35. M.J. Forrest, S.R. Holding, D. Howells and M. Eardley in *Proceedings of a Rapra Technology Conference - Silicone Elastomers 2006*, Frankfurt, Germany, 2006, 3.

36. ISO 13302, Sensory Analysis - Methods *for Assessing Modifications to the Flavour of Foodstuffs Due to Packaging*, 2003.

37. EN 1230-1, *Paper and Board Intended for Contact With Foodstuffs - Sensory Analysis - Part 1: Odour*, 2001.

38. ISO 4120, *Sensory Analysis - Methodology - Triangular Test*, 2004, 40.

39. ISO 5492, *Sensory Analysis – Vocabulary*, 2005.

40. D. van Deventer and P. Mallikarjunan, *Innovative Food Science Emerging Technology*, 2002, 3, 1, 93.

41. M. Frank, U. Ulmer, J. Ruiz, P. Visani and U. Weimar, *Analytica Chimica Acta*, 2001, 431, 1, 11.

42. C. Brede, I. Skjevrak and P. Fjeldal, *Colour Substances in Food Packaging Materials*, SNT Arbeidsrapport, Oslo, Norway, 2003.

43. C. Brede, I. Skjevrak and H. Herikstad, *Journal of Chromatography A*, 2003, 983, 1-2, 35.

44. L. Castle in Proceedings of the PIRA conference, *Plastics and Polymers in Contact with Foodstuffs*, Coventry, UK, 2003.

45. K. Bouma and E. Wijma, *Migration of Primary Aromatic Amines from Multilayer Films for Food Packaging*, The Netherlands Inspectorate for Health Protection and Veterinary Public Health, Report No. ND1FC004/01, 2002.

46. N. Palibroda, J. Brandsh, O. Piringer and R. Brandsh, *Journal of Mass Spectrometry, Letters*, 2004, **39**, 12, 1484.

47. J.W. Gramshaw and H.J. Vandenburg, *Food Additives and Contaminants*, 1995, **12**, 2, 223.

48. MAFF, *Survey of Benzene in Food Contact Plastics*, Food Surveillance Information Sheet, MAFF, London, UK, 1994, 35.

49. S.L. Herlihy, B. Rowatt and R.S. Davidson, *RadTech Europe Papers of the Month – August 2004*, RadTech Europe, The Hague, Netherlands, 2004.

50. J.W. Carter, M.S. Davis and M.J. Jupina in *Proceedings of RadTech '96 North America Conference*, Nashville, TN, USA, 1996, **1**, 29.

51. W.A.C. Anderson and L. Castle, *Food Additives and Contaminants*, 2003, **20**, 6, 607.

52. *The EFSA Journal*, 2005, **293**, 1.

53. J.A. Brotons, M.F. Olea-Serrano, M. Villalobos, V. Pedraza and N. Olea, *Environmental Health Perspectives*, 1995, **103**, 6, 608.

54. MAFF, *Survey of BADGE Epoxy Monomer in Canned Foods*, Food Surveillance Information Sheet, October, MAFF, London, UK, 1997, p.125.

55. S.R. Howe, L. Borodinsky and R.S. Lyon, *Journal of Coatings Technology*, 1998, **70**, 877, 69.

56. M. Henriks-Eckerman and T. Laijoki, *Analyst*, 1988, **113**, 2, 239.

57. C. Simoneau, A. Theobald, P. Roncari, P. Hannaert and E. Anklam, *Food Additives and Contaminants*, 2002, **19**, Supplement, 73.

58. J.K. Gillham in *Structural Adhesives: Developments in Resins and Primers*, Ed., A.J. Kinloch, Elsevier Applied Science Publishers Ltd., Barking, UK, 1986, p.1.

59. K. Dusek and I. Havlicek, *Progress in Organic Coatings*, 1993, **22**, 1-4, 145.

60. A.H. Windle in *Polymer Permeability*, Ed., J. Comyn, Elsevier Applied Science, London, UK, 1985.

61. A. Schaefer and T.J. Simat, *Food Additives and Contaminants*, 2004, **21**, 4, 390.

62. R.S. Garcia, P.P. Losada and C.P. Lamela, *Chromatographia*, 2003, **58**, 5-6, 337.

63. N. Leepipatpiboon, O. Sae-Khow and S. Jayanta, *Journal of Chromatography, A*, 2005, **1073**, 1-2, 331.

64. I. Jordakova, J. Dobias, M. Voldrich and J. Poustka, *Czech Journal of Food Sciences*, 2003, **21**, 3, 85.

65. H. Ohno, M. Suzuki, T. Aomaya and K. Mitani, *Shokuhin Eiseigaku Zasshi*, 2003, **44**, 6, 332.

66. MAFF, *Survey of Chemical Migration from Can Coatings into Food and Beverages*, 2. Epichlorohydrin, Food Surveillance Information Sheet No. 170, MAFF, London, UK, 1999.

67. *ENDS Report*, 2002, **331**, 31.

68. M. Reilly, *New Scientist*, 2007, **194**, 2603, 28.

69. M. Huber, J. Ruiz and F. Chastellain, *Food Additives and Contaminants*, 2002, **19** Supplement 1, 221.

70. B. Aurela, T. Ohra-Aho and L. Soderhjelm, *Packaging Technology and Science*, 2001, **14**, 2, 71.

71. I. Skjevrak, C. Brede, I-L. Steffensen, A. Mikalsen, J. Alexander, P. Fjeldal and H. Herikstad, *Food Additives and Contaminants*, 2005, **22**, 10, 1012.

72. H. Gaube and J. Ohlemacher in the *Proceedings of Radtech Europe - Creating Tomorrow's Technology Conference*, Edinburgh, UK, 1991, 36, 461.

73. *Plastic Packaging Innovation News*, 2006, **2**, 9, 8.

74. *Plastics Additives and Compounding*, 2004, **6**, 6, 11.

75. *Paper Film and Foil Converter*, 2005, **79**, 4, 12.

76. R.M. Podhajny, *Paper Film and Foil Converter*, 2003, **77**, 2, 22.

77. R.M. Podhajny, *Paper Film and Foil Converter*, 2002, **76**, 8, 18.

78. *Package Print and Converting International*, 2006, May-June, 8.

79. A. Lin, *Adhesives Age*, 2002, **45**, 11, 22.

80. K. Sime, *Polymers Paint Colour Journal*, 2002, **192**, 4454, 21.

81. A. Lin; G. Wind and F. Wornick in *Proceedings of the 2003 PLACE Conference and the Global Hot Melt Symposium*, Orlando, FL, USA, 2003, p.27.

82. N. Anyadike, *Flexible*, 2004, **2**, 6, 42.

83. O. Tonnoir, *European Coatings Journal*, 2004, **10**, 31.

84. R. Golden and M. Marrapese in *Proceedings of the 2004 PLACE Conference*, Indianapolis, IN, USA, 2004, p.58.

85. N. Ivory, *Polymers Paint Colour Journal*, 2004, **194**, 4482, 15.

86. *PETplanet insider*, 2005, **6**, 9, 12.

87. *Japan Chemical Week*, 2005, **46**, 2305, 2.

88. G. Cushing and L. Ostness in the *Proceedings of the 2004 PLACE Conference*, Indianapolis, IN, USA, 2004, p.54.

89. *Brand*, 2005, **4**, 4, 36.

90. J.A. Grande, *Plastics Technology*, 2005, **51**, 8, 52.

91. S. Zeren, A. Preuss and B. König, *Paint and Coatings Industry*, 2005, **21**, 4, 38.

92. Y-M. Kim, *Packaging Technology and Science*, 2002, **15**, 5, 247.

93. C.H. Lee, D.S. An, H.J. Park and D.S. Lee, *Packaging Technology and Science*, 2003, **16**, 3, .99.

94. A. Gergely in the *Proceedings of the Rapra Technology 11th International Plastics Additives and Modifiers Conference – Addcon World 2005*, Hamburg, Germany, 2005, p.12.

95. S.R. Sabreen in the *Proceedings of the 63rd SPE Annual conference – ANTEC 2005*, Boston, MA, USA.

96. S. Ver-Bruggen, *Brand*, 2004, **3**, 4, 22.

97. N. Anyadike, *Flexible*, 2003, **2**, 2, 34.

98. N. Anyadike, *Flexible*, 2004, **2**, 5, 40.

Silicone Products for Food Contact Applications

Silicone Products for Food Contact Applications

1 Introduction

The objective of the FSA Silicone food contact materials project (Contract A03046) was to provide detailed information on the types and composition of silicone based products that are used in contact with food and identify the extent to which the migration of specific constituents into food could occur. It built on information previously obtained on silicone food contact materials, such as seals and tubing, by routes such as the MAFF Rubber contact rubber project FS2219.

Silicones are used in a variety of different food contact situations and conditions. The silicone class of polymer is very versatile and the physical form of a silicone product can vary from relatively low molecular weight lubricants and oils, through high molecular weight rubbery polymers to extensively crosslinked hard resins. At the time of writing, there is no specific harmonised legislation for food contact silicone materials. However, they are covered by Regulation (EC) No 1935/2004 and a Council of Europe Resolution on Silicones (Resolution AP 2004) but this latter document is only intended to provide guidance to industry and is not legally binding.

As the breadth and scope of this review is greater than the FSA project and, because of the limitations of this format, it has only been possible to summarise the extensive amount of information that was obtained during the course of it. The reader is therefore recommended to apply to the FSA Library at 125 Kingsway, London for a full version of the final report, which was published in January 2005.

2 Silicone Products for Food Contact Applications

2.1 Silicone Polymers – Chemistry, Structure and Properties

2.1.1 Definition of a Silicone Polymer

The term 'silicone' denotes a molecule, usually polymeric and having chain segments of the general form $(R_nSiO_{(4-n)/2})_m$ where n is between 0 and 3, and m is 2 or higher.

Hence, the molecule contains a backbone of alternating silicon-oxygen atoms, with organic groups, R, attached to the silicon.

A simplified notation enables the various combinations of structural elements that are present in silicone products to be distinguished. The system that is commonly used for the polyorganosiloxanes is given in **Table 1**.

Table 1. Structural units in polyorganosiloxanes [1]			
Structural formula	**Composition**[a]	**Functionality**	**Symbol**
$R_3Si–O–$	$R_3SiO_{1/2}$	Monofunctional	M
$–O–SiR_2–O–$	$R_2SiO_{2/2}$	Difunctional	D
$RSi(–O–)_3$	$RSiO_{3/2}$	Trifunctional	T
$Si(–O–)_4$	$SiO_{4/2}$	Tetrafunctional	Q
[a] *The composition of a siloxane unit is determined by the fact that each silicon has a half share of the available oxygens, i.e., a half share of 1, 2, 3 or 4 oxygens in the above structural elements.*			

In this convention, which is used throughout this report, a linear siloxane such as octamethyltrisiloxane ($Me_3SiO–Si(Me)_2O–OSiMe_3$) is represented as MDM, and a cyclic siloxane such as octamethylcyclotetra-siloxane as D_4.

These silicone materials can be liquids or solids and possess a wide range of characteristics. They are oligomeric or polymeric (i.e., m >>2) and are man-made polymers that do not occur naturally.

Examples of silicone polymers include:

• Rubbers with an unusually wide operating temperature range extending from the very cold to very hot.

• Highly mobile fluids which do not solidify even under extreme pressures.

• Liquids of low surface tension able to form films on almost any surface.

These features, together with their high hydrophobicity, dielectric performance and release properties, combine to make silicones suitable for a diversity of applications. All are consequences of their unique molecular structure.

Table 2. Silicone Products used in food contact applications	
Product	Examples of Application
Silicone liquids	Additives to plastics and coatings, release agents for moulding plastic and rubber articles, textile impregnation agents and anti-foaming agents. Emulsifying agents in food production.
Silicone greases and pastes	Lubricants for processing machinery.
Silicone rubbers*	Baking tray coatings, baby soothers and feeding teats, food transportation tubing, valves, gaskets, stoppers, sweet moulds and spatulas.
Silicone resins	Heat resistant oven linings, non-stick heat resistant coatings on bake ware and release coatings on papers.
Silicone adhesives	Pressure sensitive adhesives for labels.
*These can be liquids (one and two part systems) that cure at room temperature or at elevated temperatures, or high MW gum rubbers.	

2.1.1.1 Summary of Principal Food Contact Silicone Products

The principal silicone products that are used in food contact applications are summarised in the **Table 2**.

2.1.2 Chemical Bonding in Silicones

In the simplest of silicones (i.e., the unmodified and non co-polymerised), the polydimethylsiloxanes (PDMS), there are only three types of bond to consider: Si–O, Si–C and C–H.

The Si–O interatomic distance in polymethylsiloxanes is 1.64 Å, which is some 5% smaller than the sum of either the atomic or ionic radii for Si and O. This, and the fact that the –O– bond angle of about 130° is much larger than the tetrahedral angle, point to significant double bond character between silicon and oxygen. The driving force to double bond formation is the difference in electronegativities between Si and O, and accessible $3d$ orbitals on the silicon allow for $(p \rightarrow d)$ dative bonds from the oxygen. The existence of a p_π–d_π bond and the geometry of the molecules both play an important role in the chemical nature of silicones. This open-chain structure, the ease of rotation around the Si–O bond, and the large size of the silicon atoms - which puts attached organic groups (e.g., methyl in polymethylsiloxanes) at a relatively large distance from the Si–O–Si backbone - allows for great molecular flexibility and a wide range of possible configurations.

Whilst the Si–C bond in PDMS is of the length expected for a normal covalent bond, the reduced steric interferences resulting from this backbone geometry also allow for good freedom of rotation around the Si–C bond. Evidence for this is seen in proton nuclear magnetic resonance (NMR) studies, which reveal unusually high mobility for the methyl hydrogens even down to very low temperatures [2]. Some polarity ($Si^{\delta+}-C^{\delta-}$) is to be expected as a result of the small difference in the respective electronegativities, but the susceptibility to attack by polar reagents is not as great as that for the more reactive Si–O bond (see Section 2.1.4).

All the C–H bonds in PDMS are primary which is a feature not duplicated in any other elastomers. Given that the ease of radical abstraction is: tertiary > secondary > primary, this bodes well for oxidative stability, and the fact that the upper temperature limit for service in dry air is around 200 °C is a reflection of this. When oxidation does occur, the products (for polymethylsiloxanes) are formaldehyde and formic acid.

2.1.3 Physical Characteristics

Linear PDMS molecules are characterised by great flexibility. This flexibility is limited only by the small steric hindrance of methyl groups: the 'forbidden' angle for movement only amounts to about 20% of the total arc. The various consequences of this flexibility include:

a) The lowest glass transition temperature (T_g is -120 °C for PDMS) of any elastomer,

b) Exceptionally low intermolecular forces resulting in:

   • Fluids of low boiling point and low viscosity (for their MW) and low temperature dependency of viscosity

   • Low surface tensions (approximately 0.02 N/m), and

   • Materials of low cohesive strength and high spreading ability.

Whilst their low surface tension means that silicones will readily wet almost any surface, a silicone surface is not easily wetted by water (γ = approximately 0.72 N/m). Silicones are characteristically hydrophobic, a property, which is used in water repellence treatments for textiles.

The solubility parameter for PDMS is 15.5 MPa$^{0.5}$, a value lower than that for polyethylene (17.0 MPa$^{0.5}$), but not quite as low as that for polytetrafluoroethylene (PTFE) (13 MPa$^{0.5}$) [3]. This low cohesive energy density means that silicones can readily take up small molecules of similar polarity. Oils and solvents rapidly swell silicone rubbers, and silicone surfaces can readily pick up organic contamination, which should be borne in mind when considering migratable components from products that have been in service.

The low surface tension of silicones is also exploited in various surfactant roles. In effect, it is the ability of the silicone to lower the surface tensions of liquids into which they are solubilised that is influential here. At very low concentrations, where the silicone is soluble in water, the reduction of surface tension helps to promote foaming (i.e., silicones are potential foaming agents). However, beyond the limit of solubility, the silicone accumulates at the water surface to reduce surface elasticity and suppress foaming. Given the low solubility of silicones in water, this defoaming action is the more widely recognised role.

2.1.4 Chemical Properties

The silicon atom in polysiloxanes is electrophilic by virtue of its empty $3d$ orbitals and so siloxanes are susceptible to nucleophilic attack. The lone pairs on the siloxane oxygens are also sites for attack by electrophiles, and the Si–O bond is susceptible to cleavage by acids or bases. The reactions take the form:

$$\equiv Si-O-Si\equiv + HX \rightarrow \equiv Si-OH + X-Si\equiv$$

As the reagents which can bring about this cleavage are also the catalysts for polymerisation, which effect predominates depends on the conditions – particularly the temperature [4]. The MW of PDMS is therefore steadily reduced by exposure to steam at temperatures in excess of 200 °C as a result of cleavage of the Si-O-Si bonds.

Whilst silicones are relatively resistant to sodium and potassium hydroxide, and provide some resistance to dilute mineral acids e.g., 10% hydrochloric acid (HCl), nitric acid (HNO$_3$) or sulfuric acid (H$_2$SO$_4$), they are unsuitable for use with any of the following:

- Ammonium hydroxide (35%)

- HCl (36%)

- Hydrofluoric acid (40%)

- HNO$_3$ (70%)

- H$_2$SO$_4$ (70%)

Two degradation routes are possible at elevated temperatures in service:

- Thermal depolymerisation (leading to lower MW cyclic polysiloxanes).

- Thermo-oxidation of the side-chain organic (e.g., methyl) groups.

Whilst the former will result in a reduction in MW, the latter may cause an increase in MW due to radical coupling (i.e., crosslinking). For example, when silicone elastomers are subjected to hot air ageing at 200 °C, crosslinking predominates and the rubbers ultimately become hard. Given that formaldehyde and formic acid are by-products of thermo-oxidative attack, and lengthy post-cures are not uncommon, the presence of such impurities in moulded silicone elastomers is a possibility.

Whilst significant thermal depolymerisation normally requires temperatures in excess of those for thermo-oxidative attack, this mode of degradation is amenable to catalysis. It can be catalysed by alkali metals, mineral and Lewis acids and various metal ions (Zn^{2+}, Cu^{2+}, and so on) [4]. Care must therefore be taken in compounding to avoid introducing impurities, which might reduce MW and introduce unacceptably high levels of migratable oligomers.

2.2 Food Contact Silicone Products – Manufacture and Composition

2.2.1 Introduction

This section covers the various types of silicone product (i.e., rubbers, fluids, resins) that are used in contact with food. For each class of product, it also describes the manufacturing processes and methods that are used in their production, and the types of additives that are used in them. These are both important as they have an influence on the species within the products that have the potential to migrate into food. A full description of these species and their migratory behaviour is provided in Section 6.

2.2.2 Manufacture of Silicone Polymers and Their Precursors

Of the different silicone product types, it is the unmodified PDMS, which have market dominance, and the discussion here will therefore concentrate on these materials. The immediate precursor to the various polymeric products is a mix of linear and cyclic polysiloxanes obtained by hydrolysis of methylchlorosilanes. The dominant route to methylchorosilanes is the so-called 'direct process' from silicon and methyl chloride [5]. The reaction involves high temperatures (e.g., at least

250 °C in the presence of a copper catalyst) and yields a complex mix of products [6]. The most abundant is dichlorodimethylsilane as represented by the equation:

$$2CH_3Cl + Si \rightarrow (CH_3)_2SiCl_2$$

The product mix is highly sensitive to temperature, for example at 400 °C, the dominant product is CH_3SiCl_3 [6]. Under normal manufacturing conditions the reaction is optimised for dichlorodimethylsilane, but a careful work-up is required to separate the different chloromethylsilanes which differ in boiling point by as little as 4 °C. **Table 3** gives an approximate product distribution for the 250-300 °C reaction.

Table 3. Approximate composition of crude chlorosilane mix from direct reaction of methyl chloride and silicon [7]	
Species	**Composition (%)**
Dimethyldichlorosilane (Me_2SiCl_2)	75
Methyltrichlorosilane ($MeSiCl_3$)	10
Trimethylchlorosilane (Me_3SiCl)	4
Methyldichlorosilane ($MeHSiCl_2$)	6
Silicon tetrachloride ($SiCl_4$)	Small amounts
Tetramethylsilane ($SiMe_4$)	Small amounts
Trichlorosilane ($HSiCl_3$)	Small amounts

The basis for polysiloxane production is that the silanols initially generated by hydrolysis of chlorosilanes are unstable in the presence of traces of acid or alkali and undergo self condensation [4], for example:

$$2Cl–Si\equiv + 2H_2O \rightarrow 2HCl + 2[HO–Si\equiv] \rightarrow \equiv Si–O–Si\equiv + H_2O$$

When the starting material is dichlorodimethylsilane, this self-condensation generates the unstable disilanol, $Me_2Si(OH)_2$. Since this is difunctional in OH, the self-condensation is potentially polymerising. In this case, both linear and cyclic polymers are obtained:

$$Me_2SiCl_2 + H_2O \rightarrow HO(Me_2SiO)_nMe_2SiOH + (Me_2SiO)_m$$

It is worth noting that the intermediate, $Me_2Si(OH)_2$ has been isolated and characterised (a solid, melting point 101 °C) - but only under strictly neutral conditions (i.e., free of all traces or acid or alkali) [8]. For a given organic grouping, the stability of organosilanols falls steeply in the order:

$$R_3SiOH > R_2Si(OH)_2 > RSi(OH)_3$$

The nature of the polymerisation on hydrolysis of Me_2SiCl_2 depends on the level of trifunctional ($MeSiCl_3$) material. This must be kept low for the manufacture of fluids or gums, when the initially formed hydrolysis products are a mixture of silanol-terminated linear PDMS and cyclic PDMS [4], represented by HOD_nMe_2SiOH and D_m, respectively. Adding some trichloromethylsilane ($MeSiCl_3$) or tetrachlorosilane ($SiCl_4$) to the dichlorodimethylsilane generates T or Q units in the siloxane backbone, and branched polymers result. Increased loadings and/or higher conversions lead to crosslinking.

The hydrolysis is autocatalytic by virtue of the HCl produced, although the balance between the cyclic and linear polymers can be influenced by added catalysis [4]. This balance can also be influenced by the presence of solvents, which allow the polymerisation to occur either in the organic phase or interfacially. Frequently used solvents include, toluene, diethyl ether, dibutyl ether and trichloroethylene. Solvents are particularly useful for viscosity control with the potentially crosslinking (i.e., higher functionality) systems that are used for silicone resin production.

2.2.3 Silicone Fluids and Silicone Gums

Dimethylsilicone fluids are made by reacting the dimethylsilicone stock (linear plus cyclic mix) with hexamethyldisiloxane [7, 9]. The reaction generates a mix of cyclic polymers and trimethylsiloxy-terminated (Me_3SiO-) linear polymers by a catalysed Si–O exchange reaction. The next equation indicates the two types of polymeric product starting with the cyclic tetramer.

$$MM + D_4 \rightarrow MD_nM + D_m$$

The ratio of MM to D_n in the charge determines the MW of the products. For relatively low viscosity fluids, an acid catalyst (e.g., H_2SO_4) is used, but for higher viscosity fluids or gums, alkaline catalysts are used. Stabilisation of the equilibrated product requires removal of the catalyst. Acid catalysts are removed by washing with water, whilst alkalies can be neutralised (e.g., with phosphite esters or carbon dioxide). Alternatively, quaternary ammonium or phosphonium salts provide alkaline catalysts, which can be decomposed thermally at the end of reaction (useful for continuous processes). After removal of the catalyst, the polymer is heated under vacuum to remove the low MW volatile products.

The resulting product mix will include:

- Various MW of trimethylsiloxy-terminated PDMS, MD_nM.

- Various MW of cyclic PDMS, D_m.

- High MW silanol-terminated linear PDMS, $HO-(SiMe_2O)_p-H$.

- Linear PDMS with both trimethylsiloxy and silanol terminals.

Silicone fluids typically have MW in the range 4,000 – 25,000 [5]. This range covers kinematic viscosities in the range 0.0001-0.001m^2/s, although viscosities as high as 0.01m^2/s can be obtained if the MW range extends to 100,000 [7, 9]. Also referred to as silicone oils, silicone fluids are exploited for their heat resistance, dielectric strength, water repellency and film-forming properties. They possess low viscosities for their MW, and their capability to remain mobile under pressures where hydrocarbon oils solidify makes them valuable hydraulic fluids [10].

2.2.4 Silicone Rubbers – from High MW Gums

Silicone rubber products for tubing, cables and seals are most commonly obtained from silicone gums [7, 9, 11]. Typical MW are in the region of 300,000 – 700,000, and conventional rubber compounding techniques are used to convert the gums into vulcanised products. Commonly used reinforcing agents are finely divided silicas (fumed silica for optimum reinforcement) although carbon black offers moderate reinforcement. Extenders (non-reinforcing fillers) are also used for cost reduction, examples include: minerals such as ground silicas, clays and calcium carbonate, and various metal oxides (e.g., iron, titanium and zinc).

In principle, plasticisers can be used to control softness, although these would not normally be used in rubbers intended for high temperature service. Nevertheless, they can be used to aid processing with the most reinforcing silicas but, usually, only silicone oils are used.

Vulcanisation of gum silicones is usually carried out using peroxides, although a catalysed silane addition cure is becoming increasingly popular. Compared with other rubbers, silicones show only low reactivities towards peroxide cures, and therefore one of the more active peroxides is essential. Some of the peroxides used in silicone cures are listed, along with their half-life data, in **Table 4**.

All of these peroxides are approved for food contact use, and their breakdown products can have a significant presence in the product vulcanisates (see Section 6).

Table 4. Organic peroxides for silicone rubber vulcanisation [11]	
Peroxide	Temperature, °C, for half life = 1 minute[a]
Bis(2,4-dichlorobenzoyl) peroxide	112
Dibenzoyl peroxide	133
Dicumyl peroxide	171
2,5-*Bis*(*t*-butylperoxy)-2,5-dimethylhexane	179
Di-*t*-butyl peroxide	193
[a] The lower this temperature the more active the peroxide.	

All these peroxides are activated by heat, giving the acronym HTV (high temperature vulcanising) to this branch of silicone rubber technology. Solid silicone HTV rubbers are amenable to vulcanisation by all the established techniques in rubber technology – such as compression, transfer and injection moulding, extrusion and calendering [11]. Given the presence of low MW siloxanes and peroxide by-products, an oven post-cure is usually used to remove as much volatile material as possible. Whilst this is primarily utilised to remove species which might cause porosity in service, the post-cure also serves to stabilise the properties for high temperature service. Ideally, the maximum temperature reached during the oven post-cure should be some 30 °C above the required temperature in service.

2.2.5 Silicone Rubbers – From Relatively Low MW Liquids

Silicone rubbers can also be made directly from liquid ingredients and this technology provides access to low-pressure moulding or casting techniques and eliminates the need for traditional rubber processing procedures, which are energy and time consuming, and capitally intensive. However, to make this possible, special silicones are required. These can be regarded as condensation polymer products held at an intermediate MW, with the final stage of the polymerisation held back until the point of manufacture.

Obtaining silanol-terminated PDMS is achieved by a controlled hydrolysis of higher MW polymers. This is accomplished by heating a high MW linear PDMS with water at 150-175 °C in the presence of alkaline catalysts. The resultant MW may lie in the region 10,000 – 100,000, and the silanol-terminated products are stabilised by careful adjustment of the pH to neutrality [7].

Nevertheless, the silanol terminals remain reactive to further condensation, and this can be performed to give network formation (crosslinking) by the addition of suitable reagents and catalysts. There are both two-pack and one-pack versions of this curing chemistry and they are both described in Section 2.2.5.1.

2.2.5.1 Two-pack Liquid Silicone Systems

Two-pack curing systems for silanol-terminated polymers can affect cure through condensation with either an alkoxysilane (silicate ester) or a silane itself [7]. The respective curing reactions are:

$$-SiMe_2-OH + RO-Si\equiv \rightarrow -SiMe_2-O-Si\equiv + ROH$$

$$-SiMe_2-OH + H-Si\equiv \rightarrow -SiMe_2-O-Si\equiv + H_2$$

The condensation by-products are, an alcohol and hydrogen, respectively. The latter reaction tends to be restricted to thin film treatments as the escaping hydrogen may lead to blowing (sponge formation) in thicker sections [7].

The equation for the alkoxy-condensation cure shows the reaction with a silicate ester. Commonly used silicate esters are tetrafunctional, e.g., tetraethyl silicate [$(EtO)_4Si$], allowing for rapid development of a crosslinked network. With suitable catalysis (Sn, Pb, Ti, Cr, Co, Zn), the cure can be effected at room temperature. Other possible curing agents are tetrafunctional titanates [$Ti(OR)_4$], or trifunctional aluminates [$Al(OR)_3$] [12, 13]

A typical curing formulation uses tetraethyl silicate [$(EtO)_4Si$] with dibutyltin dilaurate [7]. Tetraethyl silicate has been reported as being used in two-pack condensation systems to produce rubber moulds for foodstuffs [14].

Table 5 gives the ambient temperature cure rates for a range of organo-tin catalysed silicate cures.

The availability of rapid room temperature curing imposes a requirement to keep the components separate until the point of cure. Two-part room temperature curing liquid silicones are known by the acronym RTV-2.

If such condensation polymers are used for food applications, dibutyltin dilaurate is usually the catalyst, as dialkyltins do not share the toxicity (and concerns) of the trialkyltins, and they are regarded as having low mammalian toxicity [15]. Nevertheless, toxicity towards microorganisms is recognised, and dibutyltin dilaurate is used in poultry treatments (de-wormer). Headaches, irritation, abdominal pain, etc., may be amongst the symptoms experienced in human exposure to organotin: exposure to triethyltin has proved fatal [16].

Table 5. Tin-catalysed room temperature cures of silanol-terminated liquid silicones [12, 13]		
Curative	Catalyst	Gel time (min)
Ethyl silicate	Dibutyltin dilaurate	22
Butyl silicate	Dibutyltin dilaurate	92
Ethyl silicate	Dibutyltin dihexanoate	11
Ethyl silicate	Di-isobutyltin	142
2-Methoxyethyl silicate	Dibutyltin dilaurate	1.5
Tetrahydrofurfural silicate	Dibutyltin dilaurate	2.5
All the curatives in Table 5 are present at a concentration of 3 phr and the catalysts at 1 phr.		

A third process for curing liquid silicones involves a catalysed silane addition reaction [7, 12, 13]:

$$-OSiMe_2-CH=CH_2 + H-Si\equiv \rightarrow -OSiMe_2-CH_2-CH_2-Si\equiv$$

As an addition process, it offers advantages over the condensation cure in respect of better dimensional control (lower shrinkage) and the ability to cure thicker sections through eliminating the need to release volatile by-products. The reaction is catalysed with platinum group metals (i.e., platinum, rhodium, palladium or ruthenium) [7, 12, 13]. Most commonly, platinum is used at concentrations of around 2–5 ppm [12, 13] and the reaction requires silicones that contain vinyl groups (see Section 2.2.8).

A heated liquid silicone cure is known by the acronym LSR (sometimes just LR) and, by convention, this acronym is regarded as specific to the silane system.

2.2.5.2 One-pack Liquid Silicone Systems

In 'one-pack' systems, the silanol-terminated silicone, crosslinking agent and catalysts are supplied together, but the cure doesn't start until the mix is exposed to atmospheric moisture [7]. In this case the curing agent is an alkoxysilane, for example, methyltriacetoxysilane [MeSi(OCOMe)$_3$]. An initial hydrolysis of the curing agent initiates the curing reaction:

$$MeSi(OCOMe)_3 + 3H_2O \rightarrow MeSi(OH)_3 + 3MeCOOH$$

$$-SiMe_2-OH + HO-Si\equiv \rightarrow -SiMe_2-O-Si\equiv + H_2O$$

The involvement of atmospheric moisture means the reaction is subject to diffusion control and is therefore unsuitable for thick sections. Nevertheless, it works well for thin films, and the system provides a useful basis for nozzle-applied tube or cartridge sealants (e.g., familiar domestic products for kitchen and bathroom use). One-part room temperature curing liquid silicones are known by the acronym RTV-1.

2.2.6 Silicone Resins

Silicone resins are crosslinked products and the crosslinking mirrors the RTV elastomers in that the curing process is essentially an extension of the condensation by which the original polymer was prepared. The essential difference between resins and RTV elastomers is that the former has sites for crosslinking in the resin, whilst the latter introduces the crosslinking by means of a specific curative. Silicone resins are therefore highly branched and potentially crosslinkable systems, which are held at an intermediate state of conversion until the reaction can be completed as part of the final application or fabrication stage [5, 9]. Hence, whilst the polysiloxanes for RTV systems are difunctional in silanol, polysiloxanes for resins are polyfunctional in silanol as a consequence of branch points obtained by the introduction of T or Q units into the siloxane backbone.

In a typical process for silicone resin manufacture, the appropriate mix of chlorosilanes is dissolved in a solvent such as toluene or xylene and then stirred with water [5]. At the end of the reaction, the organic layer is separated and washed free of the acid produced, and then partially distilled to enrich the 'solids' content of the resulting solution. Whilst further heating or treatment with catalysts may be used to refine the MW distribution, the resin is often kept in this solution until the final cure is required. Such is the case for coatings or laminated products. Impregnation on a filler can provide the basis for moulding compounds [10].

All resin cures are affected with heat and a suitable catalyst (metal soaps, bases) [5]. Heat is essential for progressing a highly crosslinking cure, where the onset of vitrification causes a transition from kinetic to diffusion control [17]. On this basis the temperature of cure should be at least that of the highest temperature expected in service. Although somewhat softer (and less scratch resistant) than other thermosetting resins (epoxies, alkyds), silicone resins are valued for their heat resistance and water repellency.

2.2.7 Silicone Greases

Silicone greases are prepared by dispersing silica fillers in silicone oils [10]. Such two-component formulations rely on the bond between the oil and filler for their physical properties. This bond is not sufficiently strong to convey load bearing character, and such silicone greases normally have little role in lubrication. However, lubricating greases can be obtained when soaps such as lithium stearate are introduced into the formulation. In this way, silicone greases and soap-thickened greases can be used in all the roles of traditional greases and offer a wider operating temperature range than hydrocarbon-based systems.

2.2.8 Copolymers

Whilst dimethylsiloxane ($-SiMe_2O-$), is the most common repeating unit in silicone polymers, polymers containing other repeating units interspersed amongst the dimethylsiloxane are often encountered. Two such examples are the copolymers with vinylsiloxane or phenylsiloxane groupings.

2.2.8.1 Vinyl Silicones

Vinyl silicones are dimethylsiloxane-methylvinylsiloxane copolymers. The starting material to the co-monomer in random copolymers is dichloromethylvinylsilane, which is best obtained using the Grignard reagent, vinylmagnesium chloride, according to the following reaction [18]:

$$CH_2=CH-MgCl + MeSiCl_3 \rightarrow CH_2=CH-SiMeCl_2 + MgCl_2$$

The subsequent co-hydrolysis with dichlorodimethyl-silane generates copolymers with vinyl groups along the backbone. The extent of modification is not great: less than 0.5% of methyl groups replaced by vinyl gives a copolymer which, as a gum rubber, is much more active towards peroxide than the pure dimethylsiloxane homopolymer [11].

An alternative system described in the literature is one with vinyl groups on the chain ends [7]. Whilst its synthesis is not disclosed, it may be obtained using 1,3-divinyltetramethyldisiloxane which, when used in a ring-opening polymerisation of cyclic siloxanes, will result in a vinyl-terminated polymer being produced.

Both gum and liquid silicones containing vinyl groups can also be cured using a catalysed hydrosilation reaction [7, 11]. The reaction, which is platinum catalysed, is described by the following scheme:

$$-OSiMe_2-CH=CH_2 + H-Si\equiv \rightarrow -OSiMe_2-CH_2-CH_2-Si\equiv$$

A typical catalyst is hexachloroplatinic acid (H_2PtCl_6) or its solution in an alcohol ('Speier's catalyst') [19]. Hexachloroplatinic acid is approved for food contact use, and its role in the hydrosilation reaction is thought to involve both reactants in an intermediate platinum complex. The mechanism is complex, but is thought to involve oxidation and reduction steps on the platinum [20], with the Pt remaining bound to the polymer at the end of the reaction. Evidence for this is that both Pt–C and Pt–Si bonds have been detected by platinum extended X-ray absorption fine structure spectroscopy (EXAFS), and X-ray photoelectron spectroscopy (XPS) suggests that the trapped platinum catalyst at the end of reaction is in the +2 oxidation state.

Olefin complexes with vinyl silane are associated with the lower oxidation states of platinum, i.e., +2 or 0, and therefore Speier's catalyst, which as H_2PtCl_6 is a Pt(IV) compound, must undergo a reductive step in forming the first complex in the reaction sequence. In fact, this reduction step occurs during the initial solubilisation in alcohol where the $PtCl_4^{2-}$ ion is formed [21]. So at the start and end of reaction the platinum is predominantly in the +2 oxidation state.

The toxic effects commonly associated with platinum compounds are irritation and sensitisation [16] and divalent and tetravalent platinum compounds may produce strong allergic reactions (platinosis).

The induction period encountered with Speier's catalyst is reduced if a Pt-olefin complex is used as catalyst in its place [20]. For example a platinum complex with $CH_2=CHMe_2SiOSiMe_2CH=CH_2$ can be used ('Karstedt's' catalyst). If a Karstedt-type catalyst is used, then an inhibitor may be required for processability [22]. Inhibitors are suitably functional unsaturated species and examples include dimethyl maleate, diethyl maleate and diethyl fumarate. Such inhibitors will not form part of the cured product and therefore would be potentially extractable at the end of the reaction.

The final fate of the platinum appears to be the same, irrespective of the original form of the catalyst – i.e., the metal is bound into the product as a Pt(II) complex. In principle therefore, any leachable silicone can carry with it some bound platinum. Also, although it is possible that some extraction media may release the platinum in unbound form, no specific experimental evidence has been reported.

Other potentially extractable species include the carrier solvent (e.g., isopropanol) when Speier's catalyst is used, or the inhibitors (e.g., dimethyl maleate, diethyl maleate, and diethyl fumarate) if Karstedt's catalyst is used.

The co-agent is a low MW silicone having several Si–H groups [19]. A number of molecules characterised by structures such as $HMe_2SiO-(HMeSiO)_n-SiHMe_2$, where n = 0–4, have been described [2], but preparative routes have not been disclosed. The fact that the S–H bond is not quite as readily hydrolysable as the Si–Cl bond allows for a hydrolytic route to the monomer if temperatures and contact times (with the hydrolytic medium) are kept suitably low [4]. One research group has used ^{1}H-NMR analysis to characterise a commercial crosslinking agent [23]. It was reported to be a methylhydrogensiloxane (30 to 35%)– dimethylsiloxane copolymer (60 to 65%). The commercial formulation was made up of two parts (designated Parts A and B) which contained a vinyl copolymer of PDMS. Part A also contained the platinum catalyst and Part B the Si–H functional crosslinker.

The hydrosilation of vinylsiloxanes provides an addition (as distinct from condensation) cure, and delivers an ultimate reaction conversion which is not influenced by the efficiency of by-product removal. As a result, hydrosilation cures provide an alternative RTV-2 system that offer excellent resistance to compression set, and they are the preferred option for large section cures [11].

The platinum-catalysed addition cure is also gaining importance in silicone gum vulcanisation. Compared with the peroxide cure, it offers such benefits as: freedom from yellowing, no air inhibition, absence of curative breakdown products and no need for post-cure [14]. Nevertheless, it is expected that some residual Si–H crosslinker may remain after cure as a stoichiometric excess is used to provide a suitably rapid cure (23). This can be destroyed, if required, by a suitable period of high humidity storage (e.g., a few weeks at 40 °C and 75% relative humidity). However, experience and economics favour peroxide vulcanisation, which remains, for the moment, the dominant technology.

2.2.8.2 Phenyl Silicones

Replacement of some 5 to 10% of methyl groups by phenyl groups disrupts the regularity of the chain and lowers the crystallisation temperature to extend the lower service temperature of silicone rubbers to below –90 °C [11] The phenyl substitution can be achieved either through methylphenylsiloxane or diphenylsiloxane units, and provides both gum rubbers and fluids [9]. The presence of phenyl groups in methylsilicone oils improves the heat resistance and sometimes lowers the pour point [10].

The phenyl substituents can be introduced via the respective homopolymers, i.e., $-(MePhSiO)_n-$ or $-(Ph_2SiO)_n-$, using alkaline catalysis to effect the equilibration [9]. The monomeric precursors are dichloromethylphenylsilane and dichlorodiphenylsilane, with the latter being accessible via the 'direct process' from silicon and chlorobenzene [6]. The unsymmetrical dichloromethylphenylsilane is thought to be accessible via a Grignard route.

2.2.8.3 Fluorosilicones

Fluorosilicones have fluoroalkyl groups in place of methyl groups, the change modifying the solvation characteristics and providing rubbers with improved resistance to aromatic hydrocarbons, fuels and oils [24]. The fluoroalkyl group cannot be mounted directly on the silicon as the R_F–Si group is susceptible to hydrolysis – as, to a degree, is the R_F–CH_2–Si group (which is also thermally labile). For these reasons, development has focussed on the simplest grouping of the R_F–CH_2–CH_2–Si type, i.e., one with CF_3–CH_2–CH_2– groups on the backbone.

The introduction of the fluoralkyl group utilises a fluoralkene addition to the Si–H bond, which may be achieved thermally (250 to 300 °C) or, at lower temperatures, using peroxide catalysis [18, 24].

Established hydrolysis techniques (as discussed in the preceding sections) are used to generate gums which may include vinyl groups to assist subsequent vulcanisation. Compounding for vulcanisation, with fillers (silica) and curatives (peroxide type), follows the usual practice for non-fluorinated analogues.

2.2.9 Silicone Surfactants

Silicones, which may be PDMS or speciality block or graft copolymers are also key ingredients in surfactant products [25]. The nature of the surface activity of silicones was discussed in Section 2.1.3. The most widely exploited property is in defoaming, and silicone antifoaming agents are available in the undiluted form or as aqueous emulsions. PDMS-based systems are derived from silicone oils with additives as necessary to facilitate dispersion or enhance the foam suppression effect.

2.3 Food Contact and Food Related Applications

2.3.1 Release Agents

Silicones are widely used as release agents, notably to prevent sticking to metal surfaces. Particular examples include:

- Mould release agents (especially in the rubber industry).
- Coatings for metal bakeware.

In both cases, the applications compete with PTFE, although the range of silicone products allows for a degree of optimisation not necessarily accessible with PTFE. For example, in the case of mould release, the film forming properties of liquid silicones can be used for more accurate mould replication, and where a high degree of permanence is required, as in bakeware coatings, crosslinking resins can be used.

Medium viscosity oils (e.g., 0.01-0.02 m^2/s at 20 °C) are an ideal choice for mould release coatings [25]. They are sufficiently mobile for good mould coverage, and of sufficiently high MW to offer good release properties. The silicone oils are usually applied in emulsion form. Achieving an emulsion of a silicone in water is a major challenge, and dedicated emulsifying agents (part hydrophilic, part 'siliconophilic') or especially high loadings of conventional agents may be required [26]. Examples of 'siliconophilic' agents include anionic or cationic types containing organosilyl or organosiloxy groups. For good surface coverage (after evaporation of the carrier), the emulsion must achieve particle sizes below 2 µm and much of the knowledge is proprietary.

Emulsions for mould release are generally diluted to a concentration of 0.1 to 1% and applied from a spray gun directly onto the hot mould [25]. Where silicone release agents are used, some transfer of silicone to the rubber must be assumed. In view of the widespread use of silicone mould releases, any moulded rubber product can be regarded as being potentially silicone coated.

Silicone resins (see Section 2.2.6) are designed to crosslink on application, and generate surface coatings able to withstand continuous service at 260 to 370 °C (or even higher with speciality systems) [27]. Bakeware and oven linings are established applications. As in any surface coating technology, blending with other polymer systems is commonplace, and silicone-containing stoving enamels may not be exclusively silicone.

High levels of crosslinking present some barrier to diffusion, and it should be noted that coatings containing cured silicone resins do not suffer the over-coating problems of fluid-based systems [25]. It might be assumed therefore that siloxane migration is not an issue with properly-cured silicone resins.

2.3.2 Silicone Rubbers

Jerschow has reviewed the applications for different types of silicone rubber [14]. Those for food contact applications are summarised in **Table 6**.

2.3.3 Silicones as Additives for Polymers

Low levels of silicones are used as flow-enhancing additives in a variety of polymer-based systems. For example, in paints, their loadings may be as low as 0.01% [25]. In thermoplastic polymers, percentage levels may be used – perhaps let down from masterbatches containing up to 50% of silicone [25]. The main function of the additive is to ease the flow in processing, which reduces the cycle time and energy consumption. The technology is applicable to a range of thermoplastics, including: polypropylene, polyethylene, polystyrene and acrylonitrile-butadiene-styrene.

Table 6. Food contact applications of silicone elastomers				
Applications	Liquid Rubber		Gum Rubber	
	RTV-1	RTV-2	LSR	HTV
Baking tray coatings	×	×	×	
Industrial food dispensing valves		×		×
Baby soothers	×		×	
Baby feeding teats		×		×
Nipple shields	×		×	
Food dispensing valves in packaging	×		×	
Tubing for milking equipment				×
Milk liners				×
O-rings				×
Tubing for foodstuffs ingredients				×
Tubing for fermentation purposes			×	×
Stoppers for wine barrels			×	×
Stoppers for wine bottles		×	×	×
Baking moulds			×	×
Kitchen spatulas			×	×

Linear silicone polymers including those of the highest attainable MW are used and these are added as powders [28]. At very low dosing levels (0.4 – 1%), the effect of the silicone is to improve melt flow whilst, at higher levels (e.g., about 1 – 4%), migration effects provide for lubrication in processing and a better surface finish. The unwanted migration of low MW siloxanes is avoided by optimising the MW range for these additive grades.

As discussed in Section 2.1.3, silicones can also be used to modify surface tension and foaming behaviour. Control of cell size is critical in polyurethane foam production, and various silicone polymers have a role here [29]. Examples include: PDMS, polymethylphenyl-siloxanes and (more commonly) dedicated polysiloxane-oxyalkylene block and graft copolymers.

Silicones are used in the manufacture of most polyurethane foams, both flexible and rigid. A basic formulation for flexible foam is shown in **Table 7**.

Table 7. Typical formulation for a polyether-based flexible slabstock foam [30]	
Ingredient	**Amount (parts by weight)**
Polyether polyol (e.g., 3000-4000 MW triol)	63.0
Toluene diisocyanate (80:20 isomer ratio)	33.4
Water	2.7
Silicone surfactant	0.6
Amine catalysts	0.2
Tin catalyst	0.2
Pigments, dyestuffs	0.001 – 10
Fillers, flame retardants, etc.	1.0 – 200

In rigid foams, somewhat higher levels of silicone may be used to compensate for the higher polarity of the base polyols [28]. Silicones may be also be used in solid polyurethanes (cast elastomers) where their emulsifying role can help to prevent component separation in storage tanks and where foam control behaviour can help to suppress unwanted bubble formation arising from traces of moisture in the mix.

2.3.4 Silicones in Food Processing

Silicones can be directly used in food processing. For example, silicone surfactants may be used to suppress foaming at key processing stages, but they are not intended to have a role in the product and should be regarded as process aids as distinct from additives. It may be necessary therefore to take appropriate measures to remove their residues once their role has been accomplished. An example of this may be seen in a washing process, (e.g., cereals, fruit, vegetables), where a silicone antifoam may prove useful, but a further wash should follow to remove the residual silicone.

For antifoam performance in aqueous systems, the silicone is used in emulsion form; in organic media a liquid silicone can be used [31]. The base material for both systems is a PDMS fluid and Zotto [31] describes such an example, which is a methyl-terminated PDMS with a MW of around 15,000.

Organic media where such an antifoam may be useful include:

* Corn-oil manufacture
* Deep fat frying
* Esterification of vegetable oil

In addition, there are numerous aqueous systems, which can utilise silicone emulsions, for example:

- Fermentation
- Brine systems
- Cheese whey processing
- Chewing gum base
- Instant coffee and tea manufacture
- Fruit processing
- Jam and jelly making
- Juice processing
- Pickle processing
- Rice processing
- Sauce processing
- Soft drinks
- Sugar refining
- Syrup manufacture
- Vegetable processing
- Wine making
- Yeast processing

3 Regulations Covering the Use of Silicones With Food

3.1 Existing EU Legislation and Guideline Documents

At present there is no specific EU harmonised legislation for food contact silicone materials. However, they are covered by Regulation (EC) No 1935/2004, which in Article 3 states that 'substances should not migrate into food in quantities that would harm human health or affect the quality and characteristics of the food'. Annex 1 of this Regulation includes silicones within a list of materials and articles that will be covered by specific measures.

3.2 Council of Europe Resolution on Silicones (Resolution AP (2004))

The Council of Europe (an international organisation, separate from the European Union) has a committee of *Experts on Materials and Articles Coming into Contact with Food*, that meets under the auspices of the *Partial Agreement in the Social and Public Health Field*. Once adopted, the resolutions and supporting documents drawn up by these bodies are not legally binding, but members of the Partial Agreement, such as the UK, are expected to take note of them.

The *Silicone Resolution* defines the silicone product group as being comprised of silicone rubbers, silicone liquids, silicone pastes and silicone resins. Blends of silicone rubber with organic polymers are covered by the resolution where silicone monomer units are the predominant species by weight. Silicones that are used as food additives or processing aids (e.g., as defoamers in the manufacture of products such as wine), are not covered by this resolution, but polysiloxanes used as emulsifiers are. The resolution gives an overall migration limit of 10 mg/dm^2 of the surface area of the product or material, or 60 mg/kg of food. There are restrictions on the types of monomer that can be used to produce the silicone polymers and there is an inventory list. The inventory list accompanies the resolution as Technical Document No. 1 – List of Substances Used in the Manufacture of Silicone Used for Food Contact Applications. It contains some specific migration limits to complement the overall migration limit stated above. The resolution also contains a certified quality assurance scheme for the manufacture of silicone products for food contact use.

3.3 German Recommendation XV from the BfR

This is one of the most important food regulation documents for silicone materials within the EU trading block. It covers silicone rubbers, silicone oils and silicone resins. There are sections on acceptable starting materials and additives that may be used in processing and manufacture – both chemical types and addition levels. Separate restrictions are stated where silicone rubber is to be used for teats, dummies, nipple caps, teething rings or dental guards. Dummies and bottle teats must also comply with the requirements laid down in the Commodities Regulation (Bedarfsgegenstandeverordnung). A number of test methods that address some of the fundamental material properties are stipulated and acceptable limits stated. For example, for silicone rubbers, the amount of volatile material and total extractable material are both restricted to a maximum of 0.5%. There is also a test specified for the presence of unreacted, residual peroxides.

For rubbers there are test criteria that need to be met, e.g., no positive test for peroxides in the finished product, no more than 0.5% *w/w* volatile organics, and no more than 0.5% *w/w* extractable components.

3.4 Other National Legislation in the EU

In addition to Germany, there are only a few other countries within the EU that have specific national legislation that applies to silicone products. Same examples are covered in the rest of this section.

3.4.1 Belgium

There is a Silicone Resolution that has been drafted but it has yet to be adopted. In addition to this, Belgium regulates food contact materials by means of the Royal Arrete of 11th May 1992 on *Materials Intended for Contact with Foodstuffs*. This document has positive lists for various food contact materials.

3.4.2 Italy

Food contact materials in Italy are regulated under the Decree of 21st March 1973 on *Hygienic Requirements for Packaging, Containers and Utensils Intended to be Used in Direct Contact with Food and Substances for Personal Use*. It stipulates rules for the authorisation and control of objects intended to come into contact with food. Food contact materials must be prepared exclusively from components specifically listed in Attachment II according to material type (i.e., plastic, rubber, and paper and board and so on) and also comply with any other requirements specified (e.g., migration tests).

3.4.3 Netherlands

Food packaging materials are subject to the Decree of 1st October 1979 on *Packaging and Articles of Daily Use*. It contains a series of positive lists for different types of food packaging materials, each with each own chapter, for example:

- Plastics–I
- Paper and Board–II
- Rubber–III
- Metals–IV
- Glass–V
- Ceramics–VI
- Textiles–VII
- Regenerated Cellulose–VIII
- Wood and Cork–IX
- Coatings–X

3.4.4 United Kingdom

UK legislation on food contact materials is published as a number of Statutory Instruments which were published in 1978 and came into operation in 1979. For example, rubbers are covered by Statutory Instrument 1987 No 1523 *Materials and Articles in Contact with Foodstuffs*. This states that any food contact material should not be injurious to the health of the consumer and that any contamination should not have an adverse effect on the organoleptic properties of the food.

There are separate rules in the UK for the use of rubber products in contact with potable water. These are given in the UK water fitting bylaws scheme and include tests for the following:

- Taste
- Appearance
- Growth of aquatic micro-organisms
- Migration of substances that may be of concern to public health
- Migration of toxic metals

The test methods for these are given in British Standard 6920 [32].

3.5 The US Food and Drug Administration (FDA)

The FDA produces a Guidance for Industry document entitled '*Preparation of Food Contact Notifications and Food Additive Petitions for Food Contact Substances: Chemistry Recommendations*'. This is in addition to the *Code of Federal Regulations Volume 21, Parts 170 to 199 Food and Drugs*, which contains the FDA food contact regulations. This is published annually and silicone products are covered in a number of sections – see Section 4. In addition to the listed additives, the regulations also allow the use of prior sanctioned ingredients and substances that are generally recognised as safe (GRAS). The FDA regulations are relatively straightforward. Provided that the substances in a particular silicone product are listed as being approved, and it passes any specified migration test(s), it can be used for food use.

Because of the wide range of silicone products that are in existence, a number of sections of the FDA regulations can be of relevance, for example:

Section Number	Description
178.3570	Incidental food contact
177.2600	Rubber articles intended for repeated use
175.300	Resins and polymer coatings
175.105	Adhesives

4 Assessing the Safety of Silicone Materials and Articles for Food Applications

4.1 Fingerprinting of Potential Migrants from Silicone Products

There is often a need in food migration studies to carry out an initial, compositional fingerprint determination. The following techniques are commonly used to obtain fingerprint data on the species in silicone products that have the potential to migrate into food.

4.1.1 Multi-element Semi-quantitative Inductively Coupled Plasma Scan

A number of additives, e.g., fillers, can be detected in unknown compounds by carrying out a survey for some of the more popular elements using inductively coupled plasma scans (ICP). This technique enables a semi-quantitative determination to be performed for: aluminium, antimony, arsenic, barium, beryllium, bismuth, boron, cadmium, calcium, chromium, cobalt, copper, gallium, indium, iron, lead, lithium, magnesium, manganese, molybdenum, nickel, palladium, phosphorus, platinum, potassium, selenium, sodium, strontium, thallium, tin, titanium, vanadium, zinc and zirconium.

A typical detection limit for this technique is 10 mg/kg per element.

4.1.2 Targeting of Specific Species

4.1.2.1 Specific Elements

Where the compositional data obtained from a manufacturer shows that a product contains an additive for which a metal or element can be specifically targeted (e.g., the platinum-based catalyst in a silicone rubber) a specific, quantitative determination, of this element can be carried out using ICP.

In the case mentioned previously, a useful approach would be to prepare the silicone rubber sample in accordance with *British Pharmacopoeia Appendix XX F Silicone* [33] and then carry out an analysis for platinum using the ICP technique.

A typical detection limit for this type of work is around 1 mg/kg, depending on the element.

4.1.2.2 Residual Peroxides in Silicone Rubbers

Food contact silicone gum-type rubbers are often crosslinked using peroxides, a common example being dicumyl peroxide. It is possible to carry out a residual peroxide determination on the rubber using the experimental method specified in the German silicone regulations. Essentially, the method involves finely dividing 2 gram portions of the rubber and extracting them with dichloromethane (DCM) at room temperature. The DCM extract is then reacted with sodium iodide in glacial acetic acid and left in the dark. Water is then added and the liberated iodine titrated against 0.1 M sodium thiosulfate.

4.1.2.3 Silicone Additives

It is possible to use the fast and widely available attenuated total reflectance (ATR) infrared (IR) spectroscopy technique to determine the overall level of silicone components in certain food contact products where they are present as additives. For example, silicone is added to paper to produce release papers that can be used in the oven cooking of food and in packaging foods. This silicone can be analysed by ATR, with the principal benefit of the technique being that it can be performed directly on samples without any need for an intermediate preparation step (e.g., solvent extraction). Work by Suman has shown the technique to be capable of rapid, quantitative work on food release papers that had silicone levels in the region of 0.1 g/m² [34].

4.1.3 Identification of Low MW Potential Migrants

It is often useful to produce a qualitative or semi-quantitative fingerprint of the low MW species present in a silicone food contact product. As the analytical challenge is usually a complex one, the analyte being a complex mixture of low MW oligomers from the polymer, additives, and breakdown/reaction products from the complete formulation, the analytical techniques need to have both good separation and identification capabilities. Chromatographic techniques are therefore chosen and, to enable as wide a range of species as possible to be determined, three types are usually used:

- Headspace gas chromatography-mass spectrometry (GC-MS)

- Solvent extraction GC-MS

- Solvent extraction liquid chromatography-mass spectrometry (LC-MS)

4.1.3.1 Dynamic Headspace GC-MS

The dynamic version of GC-MS has certain advantages over the static version (e.g., greater sensitivity and the ability to work with much smaller sample sizes) and so this is the more popular version. In common with static headspace GC-MS, though, it can be applied to all the silicone products (i.e., rubbers, fluids and cured resins) and enables the low MW, volatile species to be identified. It is also possible to carry out semi-quantified work by reference to a standard, e.g., a low MW siloxane.

A typical set of experimental conditions for headspace GC-MS is:

Mass of sample: 5 mg

Dynamic headspace conditions:

Instrument: Perkin Elmer ATD 400

Desorption conditions: 200 °C for 10 minutes (Rubbers and Resins)

                                     150 °C for 10 minutes (Fluids)

Trap low temperature: 30 °C

Trap high temperature: 300 °C

Transfer line temperature: 220 °C

Desorption flow: Helium at 20 cm^3/min

GC-MS conditions:

Instrument: Perkin Elmer autosystem XL

Carrier: He at 7204 MPa

Column: Restek RTX-5MS 30 m × 0.25 mm, 0.25 µm film thickness

Column temperature: 40 °C for 5 minutes; 20 °C/minute to 300 °C held for 12 minutes

MS range: 20 – 450 Daltons every 0.4 seconds

4.1.3.2 Solvent Extraction GC-MS

The analysis of extracts from polymeric compounds, such as silicones, can be difficult as they are invariably complex mixtures. There have been a number of advances in GC-MS instrumentation in recent years. The advent of the GCxGC- time-of-flight mass spectrometry (ToFMS) has significantly improved the quality of the data that can be obtained. As the name implies, two separation columns, having completely different stationary phases, can be used in series and this, together with the ToFMS and the very powerful de-convolution software, enables the resolution and hence, identification and quantification, of components in very complex mixtures. The background to this technique and a description of how it can be used for the determination of low MW species that have the potential to migrate from food contact silicone products has been described recently by Forrest and co-workers [35].

Although GCxGC-ToFMS systems offer distinct advantages in situations where very high levels of separating power are required, it is also possible to use conventional GC-MS, or the GCxGC-ToFMS instruments in the one-dimensional mode, to fingerprint silicone products. Examples of the GC-MS analysis conditions that have proved to be useful for this type of work at Rapra are given in the next section.

Silicone Rubbers

Rubber samples are analysed as follows:

The samples are finely cut up and 0.3 g of each placed into 7 cm^3 vials with PTFE/silicone septa to which 2 cm^3 of acetone is added. Each sample is then extracted by ultrasonic agitation for 30 minutes and allowed to cool. The acetone extracts are then transferred to auto sampler vials and analysed by GC-MS under the following conditions:

Autosampler:

Autosampler	Agilent 6890N
Syringe size (µl)	10
Sample volume (µl)	1

Gas chromatograph:

Hardware Control	Agilent 6890
Equilibration Time (minutes)	1

GC column oven temperature ramp:

Rate (°C/min)	Target Temperature (°C)	Duration (min)
Initial	40	2.00
25.00	300	15.00

Transfer Line Temperature (°C)	300
Inlet Mode	Splitless
Inlet Temperature (°C)	280

GC column information:

Column Length (m)	30
Internal Diameter (μm)	250
Film Thickness (μm)	0.25
Phase	HP-5MS
Carrier Gas Type	Helium
Approximate Column Flow (cm^3/min)	1

Mass spectrometer:

Solvent Delay	210
Turn Filament off during solvent delay	Yes

Mass spectrometer settings:

Start Mass (μm)	45
End Mass (μm)	650
Acquisition Rate (spectra per second)	50
Detector Voltage (Volts)	1500
Ion Source Temperature (°C)	200
Masses to display during acquisition	Total ion chromatography (TIC)

In addition, solutions of known concentration of octamethylcyclotetrasiloxane in acetone can be prepared and analysed to provide a detection limit and to allow an approximate quantification of each detected component.

Silicone Fluids

Silicone fluids tend to be soluble in the majority of organic solvents, particularly non-polar solvents. However, dissolving them would result in a solution that contained species that were mostly outside the MW range of a GC-MS and of little interest in food contact work as they would be above the 1000 Daltons limit. So, to prepare a fraction that is within the MW range of GC-MS, and of interest in food contact work they can be fractionated by partitioning in methanol, as described next.

A 10 cm^3 portion of each fluid is shaken with 10 cm^3 of methanol and the methanol fraction isolated and analysed by GC-MS using the conditions described previously for silicone rubbers, except that a splitless injection technique is used and the oven time at 280 °C is 5 minutes.

As would be expected, the ratio of fluid to methanol and the time of shaking the two together has an influence on the data. This been investigated as part of the FSA project and is discussed in more detail in Section 5.4.2.

Silicone Resins

If samples of a silicone resin can be obtained prior to having been applied to a substrate, providing that they can be cured under representative conditions, specimens can be prepared for screening by extraction. If the resin is already *in situ* (e.g., on the surface of a backing tray) a portion of the whole product can be extracted. Acceptable blank samples can be difficult to prepare in the latter case and, unless the dried, cured coating weight is accurately known, even semi-quantitative treatment of the data can be difficult as the ratio of resin to extractant will be unknown.

Once these practical considerations have been addressed, extraction of the low MW species using acetone for screening can be carried out in a very similar way to that used for silicone rubbers, i.e., for cured resin samples: 0.3 g is placed into a 7 cm^3 vial with PTFE/silicone septa and 2 cm^3 of acetone added. The sample is then extracted by ultrasonic agitation for 30 minutes and allowed to cool. The acetone extracts are then examined by GC-MS under the same conditions as described for the silicone rubber extracts.

4.1.3.3 Solvent Extraction LC-MS

For the thermally labile, higher MW components in a fingerprint extract, high-performance liquid chromatography (HPLC) used to be the principal analytical choice. However, HPLC has serious limitations, which mainly relate to the detectors that are available. Species can only be accurately identified and, therefore quantified, if the relevant standards are available for confirmation of peak identity. General compositional survey work on unknown mixtures is therefore extremely difficult, if not impossible, with HPLC.

In recent years, HPLC has been consigned to the fringes of analytical chemistry with the proliferation of affordable LC-MS instruments (essentially a combination of an HPLC with a mass spectrometer). The addition of a mass spectral detector has transformed the technique and it is now being used for a much wider range of applications, particularly in the analysis of unknowns. Problems still exist, as there are no commercial LC-MS libraries (unlike those available for the more mature GC-MS technique) and the analysis conditions (i.e., mobile phase, column type and mode of detection) can have a significant influence on the data obtained.

LC-MS instruments have their own unique technology. In addition to having the in-line mass spectrometer, which itself is available in a number of different forms (e.g., atmospheric chemical ionisation (APCI) and atmospheric pressure photoinitiation, they can also be fitted with in-line UV and fluorescence detectors. The LC component has the same flexibility as an HPLC in that different columns and mobile phases (provided that they are compatible with the mass spectrometer head) can be used to ensure that a good chromatographic separation is obtained. The mass spectrometer can be operated in a number of modes, but the usual configuration results in a 'soft ionisation' process, which generates the molecular ion of each species. By changing the ionisation conditions in the head, it is possible to obtain a degree of molecular fragmentation.

Some practical guidance on the preparation of silicone products for LC-MS fingerprinting work is given next.

The main difference between LC-MS and GC-MS is that an extraction solvent needs to be used that is chromatographically compatible (i.e., miscible) with the mobile phase. As it is usual to use a polar mobile phase (e.g., a mixture of methanol and water) a relatively polar extraction solvent needs to be used. It is also usual practice with LC-MS to use a solvent that can also be regarded as a food simulant, for example either fatty food (95% ethanol) or aqueous food (distilled water) simulants.

Silicone Rubbers

Silicone rubbers can be screened using the fatty food simulant 95% ethanol as follows:

A small amount (0.3 g) of each rubber sample is placed into 2 cm^3 of 95% ethanol/5% distilled water solution and mixed ultrasonically for 30 minutes. The resulting extracts are then transferred into 2 cm^3 analysis vials and the extracts analysed by LC-MS under the following conditions:

LC-MS Conditions

Instrument:	Agilent LCMS Trap
Column:	Aqua C$_{18}$, 3 μm
Column Temperature:	50 °C

281

Flow Rate:	$0.5 \ cm^3/min$
Mobile phase:	A = Water; B = Methanol
Stop time:	120 min
Gradient:	70% B held for 0.5 min, 70% B - 100% B over 4 min then held for 115.5 min
Head Unit:	APCI in positive mode*
Scan range:	100-2200 amu
Target mass:	1000 Daltons

*Useful data is only found in the positive mode – no ionisation occurs in the negative mode

Silicone Fluids

Fingerprinting work can be carried out on dilute solutions of the silicone fluids in methanol, but the data obtained using an LC-MS set up as described previously for rubbers does not offer any advantages over the use of GC-MS. In fact no additional information is provided, and the GC-MS data can be regarded as superior in some respects, e.g., the unambiguous identification of the cyclic oligomers from their characteristic 70 eV fragmentation.

Other Silicone Products

The comments made for silicone fluids can be applied to other silicone products (e.g., resins and adhesives). An additional problem with these products is that they can be more complex with regard to their composition and the lack of any commercial identification libraries for LC-MS can severely restrict the usefulness of the LC-MS obtained, i.e., peaks will be present and a value obtained for their molecular mass, but they will remain unknown. The fact that adduct ions can be formed from interactions between the extractants and the mobile phase can also cause serious problems in the identification of unknowns.

4.2 Overall Migration Tests

The aim of an overall migration test is to determine if a food contact product is suitable for a particular food contact application by quantifying the total amount of material that will migrate from it into a certain food simulant under a specific set of conditions (e.g., temperature, time and sample area).

Food simulants are preferred to food products for this type of test as they can usually be dried down after the migration test to establish the mass of material that has migrated. This is not always the case, though, with the fatty food simulant olive oil being the prime example where a more complicated extraction/chromatographic composite method has to be used.

The precise methodology of the test to be used varies according to the regulations that need to be addressed (see Section 3), as does the way of expressing the data and the limits that have to be met. It is normal for duplicate (at least) determinations to be carried out and for the data to be adjusted for blank values.

To illustrate the different approaches, a brief description of the methods given for the determination of overall migration by the FDA and Council of Europe are given next.

4.2.1 FDA Regulations for Rubbers

Test pieces are cut from the rubber to provide a known surface area (cut edges are included in the calculation) and immersed in an appropriate amount (100 cm^3) of food simulant (either hexane or distilled water). The samples are refluxed for seven hours in pre-cleaned glassware and then removed and placed into fresh simulant and refluxed for a further two hours. The test pieces are then removed and both the seven and two hour test portions evaporated separately to dryness in conditioned crucibles and the residue weighed. Blank determinations (on equivalent volumes of the food simulant used) are also performed. The blank adjusted results have to be below specified limits (expressed in mg/in^2) in order for the rubber to be acceptable for food use.

4.2.2 Council of Europe Silicone Resolution

An overall migration limit of 10 mg/m^2 of the surface area of the product or material or 60 mg/kg of food, are given. The Resolution is less prescriptive than other regulations in that the choice of food simulant and the conditions for the overall migration experiment (i.e., time and temperature) are not specifically described, but should be appropriate bearing in mind the conditions that the silicone product will see in service. There are Council of Europe Technical Documents available that can assist workers in the design and execution of migration tests.

A complicating factor with silicone products is that their versatility means that they are available in both solid and liquid forms (see Section 2.1.2), and overall migration tests have only been designed to be carried out on solid samples.

During the course of the FSA Silicones project that was carried out at Rapra, a specific in-house method to characterise the low molecular fraction in silicone fluids was developed. The methodology that was developed is described in Section 5.2.1 and the data obtained described in Section 5.4.2.

4.3 Determination of Specific Species in Food Simulants and Foods

It has already been mentioned (Section 3) that the various silicone food regulations stipulate concentration limits for certain chemical species (additives or breakdown products) within silicone products designed for food use, and that they also have specific migration limits for certain compounds that have the potential to migrate.

Specific analytical work is therefore required:

- to ensure that a food approved silicone product is fit for purpose with respect to its chemical composition

- to ensure that compounds having standard migration limits do not exceed them in food simulants or food samples prepared using appropriate contact conditions

The use of specific analysis techniques to achieve the first objective has already been covered in Section 4.1. Their use to identify and quantify a range of chemical migrants in food simulants and food products will be discussed next.

The starting point for this type of activity is usually the carrying out of representative migration tests of the type described in Section 4.2 for the determination of overall migration. Once these tests have been performed, aliquots of the simulant (or food) are removed prior to the drying down part of the overall migration experiment so that work can be carried out to determine specific migrants.

4.3.1 Determination of Specific Elements

Where the results of compositional fingerprinting work (see Sections 4.1.1 and 4.1.2) had shown the presence of particular elements in the original silicone products, these can be targeted in the food simulants by the use of mainstream techniques such as ICP to determine the extent of migration.

For example, in addition to silicon, which would be expected in all cases, the following elements are often quantified:

- Metals, such as aluminium and magnesium, that are present in common additives such as fillers and inorganic flame retardants.

- Platinum, where a silicone rubber under test is known, or suspected, of having been cured using this type of cure system.

4.3.2 Determination of Formaldehyde

Formaldehyde is known to be produced, under certain circumstances, as the result of the high temperature oxidation of the alkyl groups within silicone products. The following methodology can be used for silicone rubbers and resins with food simulants such as distilled water and 95% ethanol. This test is not usually required for silicone fluids as this class of silicone products do not tend to be used for high temperature applications.

4.3.2.1 Silicone Rubbers

Typical contact conditions used can be 10 days at 40 °C and the method used that described in CEN 13130 Part 23 – 2005 [36]. This is the method used for food contact plastics but it can usually be considered appropriate for silicon rubbers.

4.3.2.2 Silicone Resins

A 200 cm^3 aliquot of pre-warmed food stimulant is placed into the bakeware product, which has been partially immersed in a water bath set at 90 °C. Once filled with the stimulant, the bakeware is covered with tinfoil to restrict evaporation and then left for a period of two hours. The area contacted by the food simulant is noted and calculated.

The simulant is removed after the two hours test period and set aside. The bakeware products are then refilled with simulant, heated for two hours and then the simulant discarded. Finally, the bakeware products are re-filled with the simulant for a third time and heated for the two hour test period.

The first and third contact simulant is then tested for formaldehyde using the EN 3 Part 23 – 2005 method [36].

4.3.3 Determination of Low MW Species Using GC-MS and LC-MS

The use of these techniques to obtain information on potential migrants present in silicone products (i.e., fingerprinting work) has been mentioned in Section 4.1.3. This section covers their use in identifying and quantifying specific compounds in food migration samples (simulants or food).

Once a food contact test has been performed (e.g., 95% ethanol for 4 hours at 60 °C for a silicone rubber), aliquots of the migration media are removed so that work can be carried out to determine specific migrants by GCxGC-ToFMS and LC-MS [35]. Often, particularly in the case of GC-MS, it will be necessary to extract the food migration sample (with hexane for aqueous foods/simulants, and acetonitrile for olive oil) in order to partition the analyte species into a media that is compatible with the technique. This in turn means that extraction efficiency work has to be carried out using spiked samples to determine the degree of partitioning of the analyte between the two phases. In the case of LC-MS, it is more often the case that the migration media is directly compatible with the system, but extraction/partitioning can still be required in certain cases.

In addition to the silicone specific work described by Forrest [35], the advantages of using GCxGC-ToFMS and LC-MS for the analysis of food migration samples has been reported by a number of workers [37-39] and the list of applications for these instruments is growing steadily. A brief background to the two techniques is provided next.

4.3.3.1 Two Dimensional Gas Chromatography Time-of-Flight Mass Spectroscopy (GCxGC-ToFMS)

To illustrate the data that can be obtained on food contact samples using this technique a GCxGC-ToFMS chromatogram of the acetonitrile extract of olive oil that had been in contact with one of the silicone rubbers analysed during the course of the FSA project is shown in **Figure 1**.

This 3-dimensional chromatogram clearly illustrates the fractionation by volatility on one dimension (1st column - main GC column) and by polarity on the other dimension (2nd column - GC column), with the silicone oligomers that have migrated into the olive oil and extracted by the acetonitrile being enclosed by the ellipse within the chromatogram. The majority of the other peaks within the chromatogram are chemical constituents of the olive oil that have been co-extracted with the silicone oligomers by the acetonitrile.

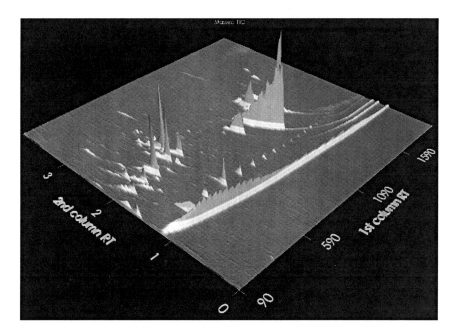

Figure 1. GCxGC-ToFMS chromatogram of the acetonitrile extract of an olive oil food migration sample from a silicone rubber sample

Analysis of Distilled Water Simulants by GCxGC-ToFMS

During the FSA project, this technique was used to identify and quantify siloxanes within a number of food contact samples (simulants and food products) that resulted from migration experiments. To illustrate the methodology used, the procedure and analytical conditions used to obtain data on distilled water samples (i.e., an aqueous food simulant) that had been in contact with silicone rubber samples is described next, starting with the conditions under which the migration experiment was conducted.

A 10 cm^3 aliquot of the distilled water simulant was placed into a pre-rinsed 40 cm^3 glass vial using a pipette and 5 cm^3 of hexane added to the vial (with a 5 cm^3 pipette). The vial was then capped and shaken for 2 minutes. The mixture was then allowed to separate before the hexane (the top layer) was removed for analysis. The same 10 cm^3 portion of aqueous stimulant was then extracted a further two times with two separate 5 cm^3 aliquots of hexane. The three extracts were then analysed separately by GCxGC-ToFMS under the following conditions, with the total amount of any siloxanes detected obtained by summing the values of the three analytical runs:

Instrument:	Agilent 6890 Gas chromatograph with LECO Pegasus III GCxGC-ToFMS
Injection:	Split injection (10:1) at 310 °C, 1 µl
Primary Column:	J and W Scientific DB-5 10 m × 0.180 mm, 0.18 µm film thickness
Secondary Column:	SGE 2 m × 0.10 mm, 0.10 µm film thickness
Carrier Gas:	Helium, 1.5 cm^3/min, constant flow
Primary oven program:	40 °C for 2 min, 10 °C/min to 320 °C. Hold for 15 min
Secondary oven program:	75 °C for 2 min, 10 °C/min to 355 °C. Hold for 15 min
Modulator offset:	40 °C
Modulator frequency:	3 s
MS conditions:	25-650 amu range scanned at 70 times/s

Semi-quantification data on the siloxane oligomers removed by the hexane from the distilled water simulant was obtained using a calibration curve produced from the cyclic siloxane tetramer.

Analysis of Siloxanes in Iso-octane Extracts of Silicone Rubber Samples

Another illustration of the specific analytical data that can be obtained on silicone products by this sophisticated form of GC-MS is shown in **Table 8**. In this case, as part of the FSA project, silicone rubbers were extracted using iso-octane and the extracts analysed by GCxGC-ToFMS.

As **Table 8** shows, upon analysis of this data, many peaks showed a reasonably consistent mass spectrum and these have been labelled as 'Siloxane' [40]. A typical mass spectrum of this type of species is shown in **Figure 2**. This spectrum is similar to the hexadecamethylheptasiloxane compound that was observed in the chromatogram eluting at a RT of ~290 seconds.

RT (s)	Name	Concentration (µg/cm^3)			
		LR3003/60	R401/60	R4001/60	R4110/60
166	Cyclopentasiloxane, decamethyl-	11.1	27.5	22.5	15.3
203	Cyclohexasiloxane, dodecamethyl-	58.7	59.9	83.0	30.6
235	Cycloheptasiloxane, tetradecamethyl-	34.2	38.4	47.9	14.7
265	Cyclooctasiloxane, hexadecamethyl-	14.7	16.1	17.7	8.0
290	Heptasiloxane, hexadecamethyl	8.6	8.5	8.3	6.1
313	Siloxane (1)	6.5	5.5	5.9	5.4
334	Siloxane (1)	6.1	4.6	5.4	5.2
353	Siloxane (1)	6.3	4.7	5.5	5.8
371	Siloxane (1)	2.7	5.3	6.8	3.5
388	Siloxane (1)	7.3	7.4	9.2	8.8
403	Siloxane (1)	12.1	11.1	12.8	12.9
418	Siloxane (1)	15.9	16.0	17.3	19.6
432	Siloxane (1)	20.1	20.9	22.1	24.1
444	Siloxane (1)	23.6	24.0	25.7	27.4
457	Siloxane (1)	24.3	25.2	24.5	28.2
473	Siloxane (1)	22.0	23.2	23.5	25.4
492	Siloxane (1)	21.9	22.9	22.6	23.3
516	Siloxane (1)	8.3	19.4	18.0	18.4
547	Siloxane (1)	14.2	14.9	14.0	7.2

Table 8. GCxGC-ToFMS data obtained on extracts from four silicone rubbers

Siloxane (1) = Higher oligomeric cyclic PDMS
RT = Retention time

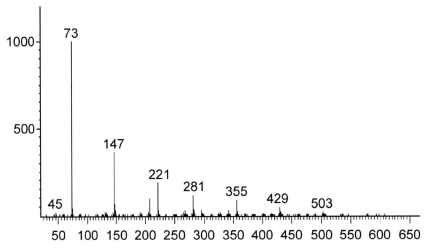

Figure 2. Typical mass spectrum for components in Table 8 labelled as 'Siloxane (1)'

4.3.3.2 Liquid Chromatography - Mass Spectrometry (LC-MS)

Mention has been made previously and in Section 4.1.3 of the advantages that LC-MS offer the analyst over HPLC. One of these advantages is that of being able to identify compounds that would not be detected using the traditional UV detector that is used in HPLC. This benefit was very apparent during the FSA project when migration samples (95% ethanol simulant) from silicone rubbers were analysed using the LC-MS in the APCI mode and the data obtained compared to that produced by the LC-MS's in-line UV detector. The results obtained are discussed next.

Analysis of 95% Ethanol Simulant Samples by LC-MS

Migration samples were produced by contacting the silicone rubbers samples with 95% ethanol. These were then analysed directly (i.e., with no intermediate extraction step) by LC-MS using a methanol-water gradient under the following instrumental conditions:

Instrument:	Agilent LC-MS Trap
Column:	Aqua C_{18}, 3 μm
Column Temperature:	50 °C
Flow Rate:	0.5 cm^3/min

Mobile phase:	A = Water; B = Methanol
Stop time:	120 min
Gradient:	70% B held for 0.5 min, 70% B - 100% B over 4 min then held for 115.5 min
Head Unit:	APCI in positive and negative mode*
MS Scan range:	100-2200 amu
In-line detector:	UV Detector

*Useful data only found in positive mode - no ionisation occurred in the negative mode.

The chromatograms obtained using these conditions are shown in **Figures 3 and 4**, with the severe limitation in the use of the traditional UV detector (the standard detector for HPLC) for studying potential migrants from silicone products highlighted by **Figure 3**, where there are no peaks present from the analysis of the 95% ethanol fatty food simulant that had been in contact with a silicone rubber. When the same simulant sample was subjected to LC-MS using the same mobile phase and column combination, a complete range of oligomer peaks were apparent (**Figure 4**).

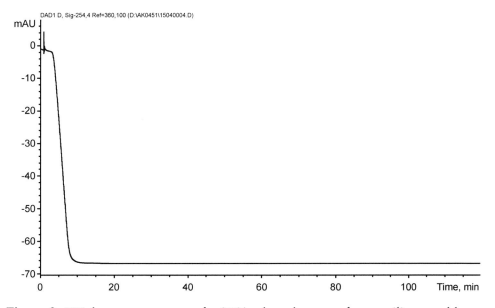

Figure 3. UV detector response of a 95% ethanol extract from a silicone rubber – no oligomer peaks are apparent

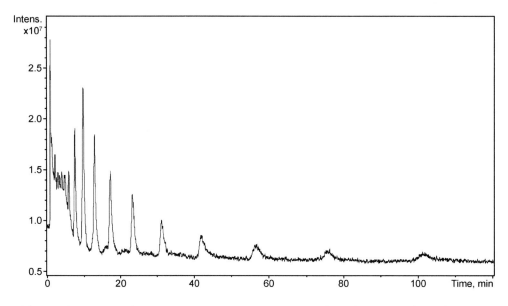

Figure 4. LC-MS chromatogram (positive APCI mode) of a 95% ethanol extract from a silicone rubber – a range of oligomer peaks is apparent

5 Foods Standards Agency Silicone Project – Contract Number A03046

The work carried out during the course of this project is summarised in this Section. However, a significant amount of data and information resulted from the work that was carried out on this project and readers are encouraged to obtain a full copy of the project report from the FSA Library, London.

5.1 Silicone Products Studied in the Project

The following, commercially available, silicone products and consumer items were obtained for this project.

5.1.1 Silicone Rubbers

a) R401/60 CI A conventional gum rubber cured using dicumyl peroxide.

b) R4110/60   A conventional gum rubber cured using a platinum catalyst.

c) R4001/60   A conventional gum rubber cured using a platinum catalyst.

d) LR3003/60 A liquid silicone rubber cured using a platinum catalyst.

These rubber samples were in the form of test sheets that had been produced by being given an initial cure and which had then been post-cured, which is the usual manufacturing procedure for silicone rubbers.

5.1.2 Silicone Fluids

a) AK1000   A PDMS fluid having a kinematic viscosity of 1000.

b) AK5000   A PDMS fluid having a kinematic viscosity of 5000.

c) AK30000   A PDMS fluid having a kinematic viscosity of 30000.

5.1.3 Silicone Resins – Uncured Products

a) MSE 100   A room temperature curing alkylsilicone resin with alkoxy groups having a low solvent content.

   Contents: Polymethoxymethyl siloxane/methylsilsesquioxane
   (CAS: 68037-85-4)
   Toluene <5.0%
   Methanol <1.0%

b) KX   A high temperature curing alkylsilicone resin in an organic solvent

   Contents: Xylene, mixed isomers >20% but <50%
   Ethyl benzene <20%
   Toluene <5%

c) M50E   A water-based (i.e., an emulsion) high temperature curing alkylsilicone resin

   Contents : α-i-tridecyl-ω-hydroxy-polyglycoether

   CAS: 9043-30-5; (surfactant) at <5%

These resins were provided in the liquid, uncured state. Some generic films of the resins were prepared at Rapra under the conditions provided by the manufacturer (e.g., 1 hour at 200 °C for the M50 E and KX products, and 2 hours at room temperature for the MSE 100 resin) for simple fingerprinting studies using techniques such as headspace GC-MS.

However, it was not possible to prepare representative, food contact quality, cured films at Rapra from these liquid resins, as the technology required [(for compounding with additives such as metallic powders, and for the spreading and curing of thin (10 μm) films)] was not available. This was a very important consideration as these resins are invariably used as binders for substances such as inert metal powders and not in a pure, undiluted form.

In order to carry out food migration testing, bakeware products that had been given a non-stick coating using a silicone-based resin were therefore sourced from local retailers (see next). The presence of a silicone non-stick coating was confirmed by both infrared analysis and an elemental analysis by x-ray fluorescence prior to any work being carried out.

5.1.4 Silicon Resin Coated Bakeware from Supermarkets

A number of examples of the following products were obtained:

a) Loaf Tin – dimensions: 21.0 cm (l) × 11.5 cm (w) × 6.5 cm (d).

b) Swiss Roll Tray – dimensions: 24.0 cm (l) × 16.0 cm (w) × 3.0 cm (d).

c) Loaf Pan – dimensions: 19.0 cm (l) × 9.5 cm (w) × 6.0 cm (d).

5.1.5 Compositional Fingerprinting Work

Compositional fingerprinting work was carried out on the products described in Section 5.1 using the range of tests described in Section 4.1. The results obtained are provided in Section 5.4.

5.2 Migration Experiments with Food Simulants

The following migration experiments were carried out on the products described in Section 5.1. The results obtained from these overall and specific migration experiments are described in Section 5.4.

5.2.1 Overall Migration Work

The following overall migration experiments were carried out on the three silicone product ranges.

5.2.1.1 Silicone Rubbers

Contact Experiments

Food simulants that represented the range of foodstuffs that silicone rubbers can encounter in service were used for this work. Silicone rubbers are typically used in high temperature applications and so relatively severe extraction conditions were chosen.

The following simulant-contact conditions combinations were used:

a) The fatty food simulant (95% ethanol) was used for 4 hours at 60 °C.

b) The aqueous food simulant (distilled water) was used for 4 hours under reflux.

c) The acidic, aqueous food simulant (3% acetic acid) was used for 4 hours under reflux – R4001/60 only.

In each case 100 cm^3 of stimulant was contacted with 0.01 m^2 of rubber. As the rubber was 2 mm in thickness, strips measuring 10 cm × 5 cm were used giving a total area of 0.01 m^2 for both sides (excluding edges).

Overall migration data was obtained by drying down the simulant extracts on a hot plate. A 20 cm^3 aliquot from each solution was removed prior to the drying operation in order to perform the specific migration tests (Section 5.2.2).

Analysis of 95% Ethanol Extracts by Gel Permeation Chromatography (GPC)

The extracts obtained by contacting the four rubbers with 95% ethanol for 4 hours at 60 °C were analysed by GPC in order to obtain their MW characteristics.

The experimental conditions used were:

A single solution of each sample was prepared by adding an appropriate amount of tetrahydrofuran to each extract to obtain a solution of 2 mg/cm^3 concentration. The solutions were mixed thoroughly and then filtered through a 0.2 μm polyamide membrane into sample vials, which were placed in an autosampler and analysed under the conditions shown next.

Columns:	PLgel guard plus, 2 × mixed bed-D, 30 cm, 5 μm
Solvent:	Tetrahydrofuran (with antioxidant)
Flow-rate:	1.0 cm^3/min (nominal)
Temperature:	30 °C
Detector:	Refractive index

Data capture and subsequent data handling was carried out using Viscotek 'Trisec 3.0' software, with calibration data obtained using polystyrene standards.

5.2.1.2 Silicone Fluids

Contact Experiments

The silicone fluids presented a problem due to their physical form. It was decided that the following procedure would provide a measure of the degree of migration occurring into simulants:

A portion (10 g) of fluid was shaken in a 100 cm³ separating funnel for 5 minutes with an amount of food stimulant (20 g). The mixture was allowed to stand and then the food stimulant removed and dried down gravimetrically. This dried extract was then analysed by GPC to obtain a MW profile of the migrant fraction.

The following food simulants were used:

- 95% Ethanol

- 10% Ethanol

- 3% Acetic acid

- Distilled water

For the 3% acetic acid and 10% ethanol simulants, 20 g of fluid was mixed with 20 g of simulant. For the 95% ethanol and distilled water simulants, the ratio was 10 g of fluid and 20 g of simulant.

It was appreciated that the amount of migration occuring could be dependent on a number of factors, such as the ratio of fluid to food simulant and the time of shaking, but this general approach enabled comparative data on the three fluids to be obtained. However, investigations were undertaken into the effect on the data of the ratio of the silicone fluid to simulant, and the contact time.

Analysis of the Food Simulant Residues by GPC

In addition to the dried residues prepared as described previously, control samples of the original AK1000, AK5000 and AK 30000 fluids were also submitted for analysis in order to obtain comparative GPC data. The conventional GPC experimental conditions used to analyse these controls and the dried residues are given next.

For the three siloxane liquid samples, a single solution of each sample was prepared by adding 10 cm^3 of chloroform to 20 mg of sample and leaving for about an hour to disperse and dissolve. The solutions were mixed thoroughly and then filtered through a 0.2 μm polyamide membrane into sample vials.

For the dried extracts obtained using 95% ethanol, solutions were similarly prepared by adding an appropriate volume of chloroform to give an approximate 2 mg/cm^3 concentration solution.

For the blank from the 95% ethanol and all the other dried extracts and blanks, solutions were prepared by adding 2 cm^3 of chloroform to residues. These were given 30 minutes (distilled water extracts), or one hour (others), to dissolve and disperse and were then transferred to the autosampler vials without filtration. This approach had to be used for these samples due to the small mass of dried extract present in each case.

The same chromatographic conditions were used as those described previously in the silicone rubbers section for the GPC analysis of the 95% ethanol samples, with the exception that chloroform was used as the mobile phase. Calibration data was obtained using polystyrene standards.

Bi-modal peak chromatograms were obtained from these conventional GPC experiments, particularly for the 95% ethanol extracts, and so one of the samples (95% ethanol extract of AK30000) was analysed by combined GPC-Fourier transform infrared (FTIR) spectroscopy to obtain qualitative data on both peaks.

The refractive index detector of the conventional system was replaced by the sample deposition module of the Laboratory Connections LC-Transform Modal 303 system and the fractionated components deposited onto a rotatable absorbent polyethylene disc. This disc was then transferred to a Nicolet Avatar 360 infrared spectrometer where spectra were recorded within the areas of the two chromatogram peaks.

Experiments to Investigate the Effect on the Data of Changing the Contact Conditions

The 95% ethanol simulant was used for this work as the GPC data had shown that this stimulant extracted the most from the fluids and that the extracts had a more obvious bi-modal distribution.

The following experiments were conducted using the lower viscosity fluids, AK1000 and AK5000:

• The proportions of the simulant and the silicone were changed from 1:2 (Fluid:Simulant) to 1:1 (Fluid:Simulant).

- The time of the extraction period was changed from five minutes, to two and eight minutes.

The extracts produced were dried down gravimetrically, as before, and then analysed by GPC.

A single solution of each sample was prepared by adding 10 cm^3 of solvent to 20 mg of sample and leaving for a minimum of four hours to dissolve. The solutions were then filtered through a 0.2 μm polyamide membrane into sample vials. Control solutions of the AK1000 and AK5000 fluids were prepared at approximate concentrations of 20, 2, 0.2 and 0.1 mg/cm^3. The same GPC conditions were used as those described previously with calibration data being obtained by analysing a range of narrow distribution polystyrene standards.

5.2.1.3 Silicone Resins

Contact Experiments

Overall migration experiments were performed on the manufactured products as follows:

Loaf Pan

This product was coated on both sides with the non-stick silicone resin and so sample strips measuring 2.5 cm × 10 cm could be cut from it and immersed in the simulants. Strips that equated to a total area of 0.01 m^2 were immersed in 100 cm^3 of simulant in each case.

The following simulant-contact conditions combinations were used:

1) 3% Acetic acid – 4 hours reflux

2) 10% Ethanol – 4 hours reflux

3) 95% Ethanol – 6 hours at 60 °C

Distilled water was not used with the resins as it was considered that purely aqueous food stuffs were unlikely to contact the resins - temperatures greater than 100 °C being the normal for bakeware. Acetic acid was however used as silicone materials can be prone to acid hydrolysis under certain conditions and it was important to investigate this. A 20 cm^3 aliquot of each solution was removed for the specific migration experiments (Section 5.2.2) prior to the gravimetric dry down procedure.

Swiss Roll Tin

This product was only coated with the silicone resin on the working surface and so migration experiments could only be conducted by filling the product with simulant. This ruled out the use of aqueous foods at reflux temperatures and so the fatty food simulant (95% ethanol) was therefore used (200 cm³) for 6 hours at a temperature of 60 °C. Again, a 20 cm³ aliquot was removed for the specific migration experiments prior to the gravimetric dry down procedure.

5.2.2 Specific Migration Work

The following specific migration tests were performed on aliquots removed from the overall migration test specimens (see Section 5.2.1).

5.2.2.1 Determination of Specific Elements

Particular elements detected by the fingerprinting work on the original silicone products were targeted in the food simulants to determine the extent of migration. Elements of interest were only detected in the LR3003/60 silicone rubber, with the following being quantified:

• Aluminium, calcium and magnesium in food simulants that had been in contact with the LR3003/60 silicone rubber sample.

• Platinum in food simulants that had been in contact with the LR3003/60 silicone rubber sample.

5.2.2.2 Determination of Formaldehyde

Silicone Rubbers and Resins

The amount of formaldehyde migrating from the silicone rubbers into the distilled water stimulant was determined using the CEN 13130 Part 23 –2005 method (36) (see Section 4.3.1). The same method was used to determine the amount of formaldehyde migrating from two of the silicone resin coated bakeware products into the distilled water and 95% ethanol simulants. The 200 cm³ of stimulant used in each case had contacted the following areas:

Loaf Pan: 0.027 m²

Swiss Roll Tray: 0.043 m²

5.2.2.3 Analysis of Food Simulants by GCxGC-ToFMS

The experimental work in this section was carried out using GCxGC–ToFMS (see Section 4.3.3)

Silicone Rubbers

- *Distilled Water Simulant*

A 10 cm^3 aliquot of the distilled water simulant was placed into a pre-rinsed 40 cm^3 glass vial using a pipette and 5 cm^3 of hexane added to the vial (with a 5 cm^3 pipette). The vial was then capped and shaken for 2 minutes. The mixture was then allowed to separate before the hexane (the top layer) was removed for analysis. The same 10 cm^3 portion of aqueous stimulant was then extracted a further two times with two separate 5 cm^3 aliquots of hexane. The three extracts were analysed separately.

The same procedure was used for all four rubber samples and the hexane extracts analysed using the following conditions:

Instrument:	Agilent 6890 Gas chromatograph with LECO Pegasus III GCxGC-ToFMS
Injection:	Split injection (10:1) at 310 °C, 1 µl
Primary Column:	J and W Scientific DB-5 10 m × 0.180 mm, 0.18 µm film thickness
Secondary Column:	SGE 2 m × 0.10 mm, 0.10 µm film thickness
Carrier Gas:	Helium, 1.5 cm^3/min, constant flow
Primary oven program:	40 °C for 2 min, 10 °C/min to 320 °C, hold for 15 min
Secondary oven program:	75 °C for 2 min, 10 °C/min to 355 °C, hold for 15 min
Modulator frequency:	3 s
Hot pulse time:	0.35 s
MS:	25-650 µm, 70 spectra/s

Semi-quantitative data on the siloxane oligomers removed by the hexane from the aqueous simulants was obtained using a calibration curve produced from a cyclic siloxane tetramer.

These experimental conditions were also used for the work on the silicone fluids and the silicone resins in this section – see next.

- *Fatty Food Simulants*

GC-MS work using a fatty food stimulant (*iso*-octane) had already been carried out on the silicone rubbers during the fingerprinting stage of the project (see Table 8) and it was decided that it was not necessary to repeat the work using this or another fatty food stimulant, e.g., 95% ethanol, as the high affinity that silicone rubbers have for this type of medium would ensure that the results obtained would be very similar, showing no important differences in terms of the types of species detected or their relative concentrations.

Silicone Fluids

- *Analysis of Separated Low MW GPC Fraction by GCxGC-ToFMS*

Analysis of the 95% ethanol simulant samples by GPC had shown that for all three fluids (AK1000, AK5000 and AK30000) two fractions were apparent – one at a relatively high MW and the other at a low MW. The low MW fraction was of particular interest because it contained species in the region (i.e., less than 1000 Daltons) that are of interest in food contact studies. The GPC system was therefore used in a preparative way to separate this fraction so that it could be characterised by GCxGC-ToFMS. As chloroform was used for the GPC work, direct analysis of the collected low MW fraction could be undertaken.

The GCxGC-ToFMS analysis conditions used were the same as those described in Section 4.3.3 for the silicone rubber samples (distilled water simulant), with the exception that the primary and secondary column oven programmes were:

Primary Column: 40 °C for 2 min, followed by 7.5 °C/min to 320 °C. Held for 15 min

Secondary Column: 75 °C for 2 min, followed by 7.5 °C/min to 355 °C. Held for 15 min

Silicone Resins

For two aqueous food simulants, a 10 cm^3 aliquot of each was placed into a pre-rinsed 40 cm^3 glass vial using a pipette. Hexane (5 cm^3) was added to the vial (with a 5 cm^3 pipette) which was then capped and shaken for 2 minutes. The mixture was then allowed to separate before the hexane (top layer) was removed for analysis. In the case of the 95% ethanol simulant samples, these were analysed directly without any work-up as they were compatible with the GCxGC-ToFMS system, the same conditions being used as those described in this Section for the silicone rubber samples.

5.3 Migration Experiments with Food Products

Migration experiments were carried out on the silicone products described in Section 5.1 using food products. The results that were obtained using these experiments are described in Section 5.4.

The following food products were used:

- **Carbonated water** – A carbonated English spring water
- **White wine** - An Italian table wine having a 11.5% alcohol content
- **Orange juice** - Orange juice made from concentrate
- **Olive oil** - Grade approved for food migration work

These four products were used to obtain specific migration data, with the carbonated water also used to obtain overall migration data. Dried down residue weights (in mg/g) were obtained for these products with the exception of the olive oil.

5.3.1 Contact Tests Performed on the Silicone Products

The silicone products were contacted with the food products under the following conditions.

5.3.1.1 Silicone Rubbers

Each of the four silicone rubbers were contacted with the following food products as follows:

Carbonated water – 4 hours under reflux (100 cm^3 used)

White wine – 4 hours under reflux (100 cm^3 used)

Olive oil – 2 hours at 175 °C (100 cm^3 used)

These conditions were chosen because they were representative of the actual use conditions of these high temperature use silicone products.

In the case of the carbonated water, a portion (40 cm^3) was removed prior to the dry down procedure to obtain a measure of overall migration. This 40 cm^3 aliquot was used for specific migration work using techniques such as GC-MS. In the case of the other two foods, no overall migration could be attempted due to a high product residue or low volatility, with the 100 cm^3 samples retained for specific migration work only.

5.3.1.2 Silicone Fluids

The following food products were contacted with the three silicone fluids:

- Orange juice
- White wine
- Olive oil

A portion of fluid (20 g) was shaken in a 100 cm^3 separating funnel for 5 minutes with 10 grams of food. The mixture was allowed to stand and then the food removed and dried down gravimetrically – except in the case of the olive oil (see next). This dried extract was then analysed by GPC to obtain a MW profile of the migrant fraction. Blanks of the dried down orange juice and white wine were also prepared and analysed by GPC using the same conditions. For the olive oil experiments, a 10 cm^3 aliquot of the oil was removed from the funnel and extracted once with 10 cm^3 of acetonitrile (again in a separating funnel) and the acetonitrile layer separated and dried down gravimetrically. This dried residue was then analysed by GPC along with the other samples using the method described in Section 5.2.1.

Calibration data was obtained by analysing a range of narrow distribution polystyrene standards. This provided MW data and the data obtained on the silicone fluid samples enabled a quantitative measure of the amount of oil in the dried extracts to be obtained.

5.3.1.3 Silicone Resins

Loaf Pan

Sample strips measuring 2.5 cm × 10 cm were cut and immersed in the food products. Strips that equated to a total area of 0.01 m^2 were immersed in 100 cm^3 of food in each case.

The following representative food-contact condition combinations were used:

- Carbonated water – 4 hours reflux

- Orange juice – 4 hours reflux

- White wine – 4 hours reflux

- Olive oil – 2 hours at 175 °C

An overall migration result was obtained for the carbonated water foodstuff but, for the other three food products, their complex nature interfered with gravimetric work and so the samples were only submitted for specific migration work.

Swiss Roll Tray

This product was only coated with the silicone resin on the working surface. Migration experiments could therefore only be conducted by filling the product with stimulant. This ruled out the use of aqueous foods at reflux temperatures. Therefore only the fatty food olive oil was used (180 cm^3) for 2 hours at a temperature of 175 °C.

5.3.2 Determination of Specific Migrants in Food Products

5.3.2.1 Determination of Silicon

Samples of the food products that had contacted the silicone rubbers and resins were analysed by ICP to quantify the total amount of silicon present. This was not carried out on the food products that had been in contact with the silicone fluids because a relatively large amount of migration occurred in these cases and the silicone oligomers present could be (approximately) quantified using techniques such as GPC.

5.3.2.2 Determination of Specific Migrants by GCxGC-ToFMS

Silicone Rubber Samples

- *Carbonated Water and White Wine*

Hexane extracts were prepared from the carbonated water and white wine migration samples and analysed as described in Section 5.2.2.

- *Olive Oil*

Sample Preparation: In each case, a 10 cm^3 aliquot of the olive oil that had contacted the rubber for two hours at 175 °C was placed into a pre-rinsed 40 cm^3 glass vial. Acetonitrile (5 cm^3) was then added to the solution. The vial was then capped and shaken for approximately 2 minutes. The mixture was allowed to separate overnight before the top layer was removed for analysis. The sample solution was extracted a further two times with a 5 cm^3 portion of acetonitrile. The same procedure was used for all the samples and the analysis conditions were the same as those described in Section 5.2.2.

Silicone Resin Samples

- *Orange Juice*

Sample Preparation: A 10 cm^3 aliquot of the orange juice that had contacted the loaf pan for four hours under reflux temperatures was placed into a pre-rinsed 40 cm^3 glass vial. Hexane (5 cm^3) was added to the solution. The vial was then capped and shaken for approximately 2 minutes. The mixture was allowed to separate before the top layer was removed for analysis. The analysis conditions were the same as those described in Section 5.2.2.

- *White Wine*

Sample Preparation: A 10 cm^3 aliquot of wine that had contacted the loaf pan for four hours under reflux temperatures was placed into a pre-rinsed 40 cm^3 glass vial. Hexane (5 cm^3) was added to the solution. The vial was then capped and shaken for approximately 2 minutes. The mixture was allowed to separate before the top layer was removed and analysed using the same conditions as those described in Section 5.2.2.

- *Olive Oil*

Sample Preparation: A 10 cm^3 aliquot of the olive oil that had contacted the loaf pan for two hours at 175 °C was placed into a pre-rinsed 40 cm^3 glass vial. Acetonitrile (10 cm^3) was added to the solution. The vial was then capped and shaken for approximately 2 minutes. The mixture was allowed to separate overnight before the top layer was removed for analysis. The sample solution was extracted for a further time with another 10 cm^3 portion of acetonitrile. The same procedure was used for the olive oil that had contacted the swiss roll tray product for two hours at 175 °C. The conditions used to analyse the extracts were the same as those described in Section 5.2.2.

5.4 Summary of Project Results

This section provides a summary of the results that were obtained on the samples prepared as described in Sections 5.1, 5.2 and 5.3. A large amount of data was generated by this project and it has not been possible to provide it all here. The results that are shown in Sections 5.4 have been chosen as they provide a representative illustration of the data that were obtained.

The compositional fingerprinting data were obtained using the analytical techniques described in Section 4.1, and the overall and specific migration data were obtained using the food simulant and food product contact conditions and analytical methods described in Sections 5.2 and 5.3.

5.4.1 Summary of the Data Obtained on the Silicone Rubber Samples

5.4.1.1 Determination of Potential Migrants – Fingerprinting Data

Cure System Species

No evidence was found for the presence of any residual peroxide breakdown products in the R401/60 C1 sample. For the platinum cured rubbers, the results shown in **Table 9** were obtained.

Elements Present

Only carbon, hydrogen, oxygen and silicone, were found at levels above 10 µg/g in R401/60, R4001/60 and R4110/60. For the liquid rubber, LR3003/60, the elements shown in **Table 10** were found.

Table 9. Platinum contents of the three platinum cured rubbers	
Rubber	**Platinum content (mg/kg)**
LR3003/60	5.0
R4001/60	1.7
R4110/60	1.1

Table 10. Elements present in the LR3003/60 rubber	
Element	**Semi-quantitative concentration**
Aluminium	13
Calcium	86
Magnesium	53

Table 11. Total siloxane oligomers present in acetone extracts of the four rubbers	
Rubber	Total mass of oligomers (µg/g)
LR3003/60	981
R4001/60	1,348
R4110/60	1,353
R401/60 C1	829

Siloxane Present

The following siloxane oligomers were identified in the four silicone rubbers by the fingerprinting work:

Trimethyl silanol (short stopper)

Cyclic oligomers (n = 3 to n >20)

Linear oligomers from n~5 to n >20

Individual masses of oligomers in an acetone extract ranged from 3 µg/g to 150 µg/g with the cyclic hexamer being the most abundant. The total siloxane oligomer content (from an acetone extract) for each rubber is shown in **Table 11**.

Overall Comments on the Fingerprinting Data

Organic peroxides are usually prolific sources of low MW species in rubber. However, prolonged oven heating will effectively purge the rubber of low MW species and given that R401/60C was a post-cured product, the absence of detectable peroxide by-products was not surprising.

The major volatile and extractable species detected in the rubbers were oligomeric PDMS, both cyclic and linear (methyl-terminated). There is a possibility that silanol-terminated oligomers were present, but no evidence for these was found.

5.4.1.2 Migration Work Using Food Simulants

Overall Migration Results

The results obtained are shown in **Table 12**.

The values in **Table 12** can be compared to the silicone resolution limit of 10 mg/dm². The aqueous results are well within this value and, although the fatty food simulant (95% ethanol) data is in excess of it, it has to be borne in mind that the conditions under which the test was conducted were relatively severe.

• *MW Characteristics of the 95% Ethanol Extract*

The GPC results obtained on these samples are summarised as in **Table 13**.

Specific Migration Results

Platinum: The distilled water and 95% ethanol simulants that were in contact with the three platinum cured rubbers were analysed for their platinum content. In all three cases the result obtained was less that the detection limit of 0.2 μg/cm³, which corresponds to <0.02 mg/dm²

Elementals – Other Than Platinum (LR3003/60 only): The data obtained on the distilled water and 95% ethanol samples are shown in **Table 14**.

Formaldehyde: Formaldehyde was determined in the distilled water simulant using CEN 13130 Part 23-2005 [33] and the results were:

R401/60, R4110/60 and R4001/60 <0.5 mg/kg
LR3003/60   ~0.5 mg/kg

Although not applicable, it can be seen that these values are much less than the 15 mg/kg limit given in the Plastics Regulations (2002/72/EC).

Silicon: The results obtained are shown in **Table 15**.

Silicone Oligomers (Distilled Water Data): The silicone oligomers present in the distilled water food simulant were determined by GCxGC-ToFMS. The types of species, and the approximate levels identified (via a hexane extraction step), in the distilled water that had been in contact with the four rubbers are shown in **Table 16**.

Table 12. Overall migration data obtained on the four rubbers			
Rubber	Overall Migration (mg/dm^2)		
	Distilled water	3% acetic acid	95% ethanol
R401/60 C1	5	-	29
R4001/60	4	2	29
R4110/60	4	-	58
LR3003/60	2	-	21

Table 13. GPC data obtained on the 95% ethanol extracts of the four rubbers		
Rubber	M_w	M_n
R401/60 CI	2,240	1,490
R4001/60	1,710	1,350
R4110/60	1,680	1,360
LR3003/60	1,950	1,410

M_w = *weight average MW*
M_n = *number average MW*

Table 14. Elements detected in food simulants that had contacted LR3003/60		
Element	Concentration (mg/dm^2)	
	Distilled water	95% ethanol
Aluminium	<0.01	<1.0
Calcium	0.05	<0.5
Magnesium	<0.01	<0.5
Silicon	5.52	4.3

Table 15. Silicon present in food simulants that had contacted the four rubbers				
Simulant	Concentration (mg/dm^2)			
	R401/60 C1	R4001/60	R4110/60	LR3003/60
Distilled Water	<1.0	<1.0	<1.0	<1.0
95% Ethanol	6.0	4.7	14.0	4.3
3% Acetic acid		<1.0		

Table 16. Silicone oligomers present in the distilled water simulant that had contacted the four rubbers				
Siloxane	R401/60 (μg/dm²)	R4110/60 (μg/dm²)	4001/60 (μg/dm²)	LR3003/60 (μg/dm²)
Cyclotetrasiloxane, octamethyl-	39	30	44	nd
Cyclopentasiloxane, decamethyl-	52	36	64	17
Cyclohexasiloxane, dodecamethyl-	43	27	67	23
Cycloheptasiloxane, tetradecamethyl-	15	nd	27	15
Cyclooctasiloxane, hexadecamethyl-	nd	nd	13	nd
nd = not detected				

Table 17. Overall migration results obtained using carbonated water	
Rubber	Overall Migration (mg/dm²)
R401/60 C1	0.9
R4110/60	7.6
R4001/60	3.2
LR3003/60	4.8

Table 18. Silicon present in food products that had contacted the four rubbers				
Food	Concentration (mg/dm²)			
	R401/60 C1	R4001/60	R4110/60	LR3003/60
Carbonated water	1.7	1.1	1.8	4.1
White wine	<1.0	0.5	0.2	0.3
Olive oil	14.5	9.3	10.7	8.4

Table 19. Silicone oligomers present in the carbonated water that had contacted the four rubbers				
Siloxane	R401/60 (μg/dm²)	R4110/60 (μg/dm²)	4001/60 (μg/dm²)	LR3003/60 (μg/dm²)
Cyclotetrasiloxane, octamethyl-	17.4	8.4	19.3	9.4
Cyclopentasiloxane, decamethyl-	14.0	11.9	18.1	9.3
Cyclohexasiloxane, dodecamethyl-	8.7	nd	9.2	8.0
Cycloheptasiloxane, tetradecamethyl-	6.9	nd	6.7	7.0
nd = not detected				

Migration Work Using Food Products

Overall Migration Results – Using Carbonated Water: The results obtained are shown in Table 17. All of the values are below the EU Silicone Resolution limit of 10 mg/dm^3.

Specific Migration Results

Silicon: The results obtained are shown in **Table 18**.

Carbonated Water: The oligomeric data obtained on the four rubber samples are shown in **Table 19**.

White Wine: Only the R4001/60 rubber sample showed an increase in the level of siloxanes compared to the control, the results obtained are shown in **Table 20**.

Olive Oil: The data obtained on the four rubber samples is shown in **Table 21**.

Table 20. Silicone oligomers present in the white wine that had contacted the four rubbers	
Peak assignment	μg/dm^2
Cyclotetrasiloxane, octamethyl-	9.7
Cyclopentasiloxane, decamethyl-	10.2
Cyclohexasiloxane, dodecamethyl-	27.3
Cycloheptasiloxane, tetradecamethyl-	13.8
Cyclooctasiloxane, hexadecamethyl-	8.1

Table 21. Silicone oligomers present in the olive oil that had contacted the four rubbers				
Siloxane	R401/60 (μg/dm^2)	R4110/60 (μg/dm^2)	4001/60 (μg/dm^2)	LR3003/60 (μg/dm^2)
Cyclotetrasiloxane, octamethyl-	1.1	0.2	0.4	nd
Cyclopentasiloxane, decamethyl-	13.5	18.8	9.0	2.8
Cyclohexasiloxane, dodecamethyl-	27.4	32.9	26.5	8.9
Cycloheptasiloxane, tetradecamethyl-	8.6	11.2	9.3	2.9
Cyclooctasiloxane, hexadecamethyl-	3.6	5.9	3.8	0.9
Higher cyclic oligomer	2.2	3.9	2.4	0.3
nd = not detected in the acetonitrile extract.				

5.4.2 Summary of the Data Obtained on the Silicone Fluids

5.4.2.1 Potential Migrants – from Fingerprint Data

The fingerprinting data obtained on the four silicone fluids were as follows.

Elemental Data

Apart from silicon, hydrogen, oxygen and carbon, no elemental species were found to be present above 20 µg/g.

Siloxanes

Characterisation work by both headspace GC-MS and solvent partition (extract) GCxGC-ToFMS showed the presence of a full range of dimethyl siloxane cyclic and linear oligomers from the trimer to >20. The relative abundance of the oligomers increased in the order AK1000 > AK5000 > AK30000. All three fluids were also found to contain the low MW species diethoxydimethoxysilane and, tentatively (from the mass spectrum match), dimethoxymethylsilane.

Overall Comment on the Fingerprint data

The ranking with respect to the relative abundance of these volatile and extractable oligomers reflects the viscosities (and hence MW) of these samples. The kinematic viscosities are given in the respective material codes (i.e., 1000 for AK1000) and thus the viscosity increases in the order AK1000 → AK5000 → AK30000.

5.4.2.2 Migration Work Using Food Simulants

Overall Migration Data

- *Extract Weights*

Representative dried extract weights for aqueous and fatty food simulants are shown in **Table 22**.

The essentially hydrophobic nature of the silicone fluids accounts for the fact that relatively small amounts of the silicone fluid, irrespective of MW, partitioned into the aqueous simulant. This is also why far more of the silicone fluids partition into the hydrophobic, fatty food simulant (95% ethanol). For this simulant, the viscosity of the fluid is important, with the amount of partitioning decreasing with increasing viscosity.

Two of the fluids (AK1000 and AK5000) were used, together with the 95% ethanol simulant, to investigate the effect of varying the ratio of silicone fluid to simulant. Experiments were performed using ratios of 1:2 and 1:1. The contact time remained at five minutes and the results obtained are shown in **Table 23**.

This showed that the amount of residue obtained is dependent on the amount of simulant used with the amount of silicone partitioning into the stimulant increasing with increasing proportion of simulant.

Table 22. Overall migration data obtained using distilled water and 95% ethanol (Ratio used: Silicone fluid 10 g/simulant 20 g)		
Silicone Fluid	Extract weight (mg/g of silicone fluid)	
	Distilled water	95% Ethanol
AK 1000	0.30	6.90
AK 5000	0.09	5.80
AK 30000	0.12	2.73

Table 23. Effect on extract weight of varying the silicone fluid:simulant ratio		
	Extract weight (mg/g of silicone fluid)	
	Ratio 1:1 (Fluid:Simulant)	Ratio 1:2 (Fluid:Simulant)
AK 1000	1.69	6.90
AK 5000	1.48	5.80

Table 24. Effect on extract weight of varying the contact time		
Silicone Fluid	Extract weights obtained (mg/g silicone fluid)	
	Two minutes	Eight minutes
AK 1000	0.73	0.81
AK 5000	0.55	0.54

The same two fluids and food stimulant were used to investigate the effect of varying the contact time – times of two and eight minutes were used with the ratio of fluid to simulant kept at 2:1. The results obtained are shown in **Table 24**.

This showed that increasing the contact time four-fold had little effect on the amount of AK5000 partitioning into the food stimulant; a greater difference being apparent with the lower viscosity silicone fluid (AK 1000) due to its greater mobility.

- *MW Characterisation of the Food Simulants Extracts by GPC*

95% Ethanol Extracts: Analysis of the 95% ethanol extracts by GPC showed that they all contained both a high and a low MW component. Representative GPC data for these two components, using a silicone fluid:simulant ratio of 1:2, is shown in **Table 25**. The GPC data obtained on each silicone fluid is included in **Table 25** for comparison.

The 95% ethanol stimulant was also used to investigate the effect of varying the ratio of fluid to simulant and the contact time. The dried extracts obtained were analysed by GPC. Representative results are shown in **Tables 26 and 27**.

For comparison, the GPC data for the extracts obtained using the aqueous food simulant distilled water are given in **Table 28**.

Specific Migration Data

- *Analysis of the Low MW fraction of the 95% Ethanol Fraction by GCxGC-ToFMS*

The low MW fraction that was present in the 95% ethanol extracts was of particular interest, a substantial part of it residing in the mw region of interest in food contact studies, and so it was separated and analysed by GCxGC-ToFMS. The data obtained showed that the fraction contained over 25 siloxane oligomers, both cyclic and linear. The major species were in the range hexamer to nonamer.

5.4.2.3 Migration Work Using Food Products

Overall Migration Data

- *Extract Weights*

The silicone fluids were contacted with a range of food products. For orange juice and white wine, the blank values were higher than the samples that had contacted

Table 25. GPC data obtained on the 95% ethanol extracts				
Sample	High MW component		Low MW component	
	M_w	M_n	M_w	M_n
AK1000 Fluid				
Siloxane fluid	36,000	21,400	1,670	1,480
95% Ethanol extract	37,400	24,100	1,190	900
AK5000 Fluid				
Siloxane fluid	62,500	33,700	1,800	1,370
95% Ethanol extract	66,200	40,000	1,130	940
AK30000 Fluid				
Siloxane fluid	104,000	54,800	1,910	1,350
95% Ethanol extract	113,000	71,600	1,160	930

Table 26. GPC data on the 95% ethanol extracts (high MW component only)				
Sample	M_w	M_n	M_w/M_n	Concentration of silicone* (mg/cm³)
AK1000 (1:1)	36,900	23,400	1.6	0.41
AK1000 (1:2)	37,400	24,100	1.6	nd
AK5000 (1:1)	69,600	39,200	1.8	0.22
AK5000 (1:2) (Fluid:Simulant)	66,200	40,000	1.7	nd

Provides a relative measure of the amount of silicone in the chromatogram peaks calculated from a calibration mass response for the silicone fluids

Table 27. GPC data on 95% ethanol extracts (low MW component only)				
Sample	M_w	M_n	M_w/M_n	Concentration of silicone* (mg/cm³)
AK1000 (1:1)	1,270	1,040	1.2	1.65
AK1000 (1:2)	1,190	900	1.3	nd
AK5000 (1:1)	1,180	990	1.2	1.62
AK5000 (1:2) (Fluid:Simulant)	1,130	940	1.2	nd

nd = not determined for these runs.
Provides a relative measure of the amount of silicone in the chromatogram peaks calculated from a calibration mass response for the silicone fluids

Table 28. GPC data obtained on distilled water extracts			
Sample	M_w	M_n	M_w/M_n
AK1000	36,100	17,000	2.1
AK5000	64,600	26,000	2.5
AK30000	102,000	27,200	3.8

Table 29. GPC data of the orange juice extracts					
Sample	M_w	M_n	M_w/M_n	Concentration of silicone (mg/cm³)	%S
AK 1000	33,700	11,500	2.9	0.64	3.2
AK 5000	55,500	9,510	5.8	0.18	0.9
AK 30000	88,800	4,190	21	0.07	0.4

%S = Approximate percentage silicone in total orange juice extract
Note: There was no response for the orange juice blank

Table 30. GPC data of the white wine extracts					
Sample	M_w	M_n	M_w/M_n	Concentration of silicone (mg/cm³)	%S
AK 1000	32,600	8,090	4.0	1.48	7.4
AK 5000	No Response				na
AK 30000	No Response				na

%S = Approximate percentage silicone in total white wine extract
na = not applicable.
Note: There was no response for the white wine blank

Table 31. GPC data of the Olive oil extracts					
Sample	M_w	M_n	M_w/M_n	Concentration of silicone (mg/cm³)	%S
Olive oil blank	1,370	1,300	1.1	2.25	na
AK 1000	1,680	920	1.8	2.20	110
AK 5000	1,550	920	1.7	2.07	104
AK 30000	1,490	890	1.7	2.16	108

%S = Approximate percentage silicone in olive oil extract
na = not applicable

the fluids. This was thought to be due to the fact that more migration occurred from these food products (particularly of components such as sugars) into the fluids than the other way round.

The olive oil was extracted with acetonitrile using a volume ratio of 1:1. Higher extract weights were obtained from the lower viscosity fluids, which indicated that more of these fluids had migrated into the olive oil.

- *MW characterisation of the food product extracts by GPC*

The dried extracts obtained from the fluid:simulant contact experiments were analysed by GPC. Representative results for the food products are shown in **Tables 28 to 31**.

The %S values are greater than 100% due to the presence of extracted components from the olive oil itself. It was not possible from the GPC data obtained to separate the silicone oligomers from these olive oil components.

- *Specific Migration Data*

The overall migration work carried out on the silicone fluids using the food simulants and food products had shown that fractionation of the fluid did not occur, the GPC data showing this migrated material to have a similar MW distribution to the original silicone fluid. As expected, the amount of a given fluid that dissolved in the food was proportional to its viscosity. The silicone fluids did not have any additives in them and were solely composed of PDMS oligomers and this restricted the options for specific migration work. Analysis of these migrating oligomers using other chromatographic techniques, such as GC-MS, was not applicable as the range of MW involved was far in excess of the capability of these techniques.

An important consideration here is that the contact tests for these products had to be improvised for this project (no standard migration test being available) and that work had shown that varying the experimental conditions could result in significant differences in the data obtained. This project data, therefore, was of value in indicating the type of migration behaviour that occurs with various food products, and how this is affected by the conditions of contact and the viscosity of the fluid, but was indicative only and should not be taken as absolute values.

5.4.3 Summary of the Data Obtained on the Silicone Resin Samples

5.4.3.1 Potential Migrants – from Fingerprinting Data

The fingerprinting data obtained on the three silicone resin samples are as follows:

Elemental

Only carbon, hydrogen, oxygen and silicon were found to be present in the MSE 100 and KX fluids above 10 µg/g.

For the emulsion-based resin, M50E, in addition to carbon, hydrogen, silicon, oxygen and potassium (at 26 µg/g) and sodium (at 270 µg/g) were also present. Both of these species relate to the surfactant stabiliser system in the emulsion.

Silicone Oligomers

Analysis of cured specimens of the three resins by both headspace and solvent extraction GCxGC-ToFMS revealed volatile and extractable species to be essentially siloxane in character, although precise identifications were often difficult due to poor library matches. Silicone resins utilise different polymers from those commonly used in silicone rubbers or fluids, and so a diversity of options was possible. Species identified in the resins included dimethylsiloxanes and methoxymethylsiloxanes and some evidence of ethoxysiloxanes.

Oxidation Products

Two of the resins (KX and M50E) were cured at elevated temperature and this resulted in some oxidation products being formed. Acetaldehyde, acetic acid, acetone, and ethanol were found.

It should be emphasised however that Rapra did not have the technology necessary to cure these resins to the same standard and condition that would be used in industry and so it cannot be guaranteed that these species would be formed during commercial production.

Overall Comments on the Fingerprinting Work

Given the abundance of the primary hydrogen in PDMS, the polymers should be relatively resistant to oxidation. When it occurs, formaldehyde and formic acid would be expected. The fact that a higher aldehyde and acid, namely acetaldehyde and acetic acid, were obtained from these resin cures suggests a source other than PDMS oxidation. It is more likely that their origins lie in the condensation crosslinking where ethanol is commonly produced, acetaldehyde and acetic acid being recognised oxidation products of ethanol.

5.4.3.2 Migration Work Using Food Simulants

Overall Migration Data

The overall migration results obtained using the loaf pan and swiss roll products are summarised in **Table 32**. The use of 3% acetic acid with the loaf pan was not successful due to rusting of the cut sample surfaces.

Table 32. Overall migration data obtained using food simulants		
Product	10% ethanol (mg/dm²)	95% ethanol (mg/dm²)
Loaf pan	1.32	0.55
Swiss roll tray	nd	1.38
nd = none detected		

Specific Migration Data

Silicon: The results obtained are shown in **Tables 33 and 34**.

Formaldehyde: The results obtained for the swiss roll tray and loaf pan bakeware products are shown in **Table 35**.

The results show the normal trend that the amount of formaldehyde detected is greater in the third contact test than the first. This is because the formaldehyde is generated as an oxidation product of the silicone resin and its concentration will increase as the heat history (degree of ageing) of the material increases.

The results shown previously can be compared to the formaldehyde limit of 15 mg/l (15 mg/kg) in the food simulant given in the plastics regulations - all of the values being very much lower than this limit.

Food Simulant	Concentration			
	1st Reflux		3rd Reflux	
	$\mu g/cm^3$	mg/dm^2	$\mu g/cm^3$	mg/dm^2
3% Acetic acid	<10	<1.0*	<10	<1.0*
10% Ethanol	100	10	<10	<1.0*
95% Ethanol	10	0.72	<10	<0.72*

Table 33. Silicon present in the food simulants that had contacted the loaf pan

*The detection limit of the original analysis (i.e., on the simulant) is 10 $\mu g/cm^3$. These values have been produced using this 10 $\mu g/cm^3$ figure but could be much lower.

Table 34. Silicon present in the food simulants that had contacted the swiss roll tray

Food Simulant	Concentration	
	$\mu g/cm^3$	mg/dm^2
95% Ethanol	<10	<0.42*

*The detection limit of the original analysis (i.e., on the simulant) is 10 $\mu g/cm^3$. These values have been produced using this 10 $\mu g/cm^3$ figure, but could be much lower.

Table 35. Formaldehyde present in the food simulants that had contacted the swiss roll tray and the loaf pan

	Formaldehyde concentration	
	1st Contact	3rd Contact
Swiss Roll Tray		
Distilled Water	0.29 mg/l (0.013 mg/dm²)	0.095 mg/l (0.004 mg/dm²)
95% Ethanol	nd	0.049 mg/l (0.002 mg/dm²)
Loaf Pan		
Distilled Water	nd	nd
95% Ethanol	nd	0.029 mg/l (0.002 mg/dm²)

Siloxane Oligomers: GCxGC-ToFMS analysis of the hexane extracts of the two aqueous food simulants (10% ethanol and 3% acetic acid), and a direct analysis of the 95% ethanol simulant were carried out. The following comments can be made after a comparison of the sample data with the control data:

1) 10% ethanol simulant: control blank siloxanes were at higher levels than either the first or second hexane extracts.

2) 95% ethanol simulant: control blank siloxanes were at slightly higher levels than either simulant sample.

3) 3% acetic acid simulant: control blank siloxanes were at higher levels than either the first or the second hexane extracts.

In conclusion, the samples have produced results which are the same, or similar, to the blanks showing that no detectable level of migration has occurred into the simulants.

5.4.3.3 Migration Work Using Food Products

Overall Migration Data

Overall migration work cannot usually be carried out on food products because they usually give large dried residues due to the sugars and other ingredients present. This experiment was worth conducting on the carbonated water, however, as it only gave a relatively small residue.

Carbonated Water: The loaf pan product was tested using this simulant for four hours at reflux temperatures and the results obtained were:

Actual sample residue weights:
 14.57 mg and 13.98 mg

Blank determination:
 14.37 mg

Blank corrected duplicate migration weights:
 0.2 mg and –0.39 mg

Result:
 Overall migration values 0.2 mg/dm^2 and –0.39 mg/dm^2

The values obtained were very small relative to the blank (which was larger than usual due to the fact that this is a food product and not a simulant) and this explains why one of the duplicates gave a negative value once the blank had been taken into account.

Specific Migration Data

- *Silicon*

The results obtained are shown in **Table 36**.

Table 36. Silicon present in the food products that had contacted the loaf pan and the swiss roll tray			
	Concentration		
	$\mu g/cm^3$	mg/dm^2	mg/kg (6 dm^2/kg)
Loaf Pan			
Carbonated water	17	1.22	7.3
Orange juice	<10	<1.0	<6.0
White wine	18	0.5	3.0
Olive oil	10	0.72	4.3
Swiss Roll Tray			
Olive oil	<10	<0.43	<0.61

None of these results showed a significant enhancement over the blank values. They indicate that very little, if any, migration of silicone oligomers occurs, from either resin, into any of the food products.

- *Silicone Oligomers*

The results obtained are:

Orange Juice: The levels of siloxanes in the hexane extract of the samples were at very similar levels to the control sample.

White Wine: The levels of siloxanes observed in the hexane extract of the white wine control and hexane extracts of the duplicate loaf pan samples were at very similar levels.

Olive Oil: The olive oil that had contacted the loaf pan and swiss roll tray products was extracted with acetonitrile (at an efficiency of 33%) and analysed by GCxGC-ToFMS. The results obtained are discussed next.

• *Loaf Pan Product*

When a comparison was made between the data obtained on the duplicate loaf pan samples and the olive oil control, no significant differences were apparent between the duplicates and the olive oil control.

• *Swiss Roll Tray Product*

With the swiss roll tray resin sample it was possible to find an enhancement of the siloxanes present in the duplicate samples relative to the control.

Representative GC-MS silicone oligomer data obtained on the swiss roll tray product is shown in **Table 37**.

Table 37. Silicone oligomers in the olive oil that had contacted the swiss roll tray				
Peak assignment	Approximate amount (ng/cm^3)*	$\mu g/dm^2$**	mg/kg (6 dm^2/kg)	mg/kg*** (6 dm^2/kg)
Cyclopentasiloxane, decamethyl-	166	3.6	0.022	0.07
Cyclohexasiloxane, dodecamethyl-	257	5.5	0.033	0.10
Cycloheptasiloxane, tetradecamethyl-	156	3.3	0.020	0.07
Cyclooctasiloxane, hexadecamethyl-	156	3.3	0.020	0.07
Total amount of silicone oligomers	620	13.2	0.079	0.26

* Amount of species observed in the acetonitrile extract of the olive oil that had contacted the swiss roll tray product.
** Specific migration value taking into account the amount of olive oil (180 cm³) and the contact area (4.2 dm²).
*** Values allowing for the 33% extraction efficiency that the acetonitrile exhibited in removing siloxanes from the olive oil.

The difference in the data found between the two resin coated products may be due to the fact that the 180 cm^3 of olive contacted a far greater area in the case of the swiss roll tray product (4.2 dm^2) than was the case with the loaf pan product (2.5 dm^2). This difference is due to the shape of the products.

5.4.4 Overall Summary of the Project and the Results Obtained

The data shown in Sections 5.4.1 to 5.4.3 are complex and to assist in its interpretation a succinct summary of the project and its findings is given next.

5.4.4.1 Overview

During the course of the project, experimental work was carried out on 10 representative commercial silicone products - four silicone rubbers, three silicone resins and three silicone fluids. Compositional fingerprinting of these products was carried out using a range of analytical techniques (elemental, GC-MS, GCxGC-ToFMS, and LC-MS) to determine any low MW species that had the potential to migrate into food. The data obtained in the initial compositional fingerprinting exercise was used as guide to plan the experimental work for migration experiments involving both food simulants and food products. The results of this fingerprinting work revealed that, although species other than silicone oligomers, such as peroxide curative breakdown products, can be predicted as potential migrants, only oligomers ranging from the trimer to n >20 were found. Of these, the smaller, cyclic oligomers (hexamers) were the most abundant.

Overall and specific migration experiments, using materials and methods for EU testing of food contact plastics, were carried out on the silicone products using food simulants (distilled water, 3% acetic acid, 10% ethanol, 95% ethanol). Specific migration experiments were also carried out using a number of food products (carbonated water, orange juice, white wine and olive oil). Analysis of the samples produced by specific migration experiments was carried out using a range of elemental (ICP), specific (for formaldehyde) and chromatographic (GCxGC-ToFMS, LC-MS and GPC) techniques.

5.4.4.2 Data Obtained for Silicone Rubbers

The food migration experiments carried out on the silicone rubbers showed that the principal species of interest were siloxane oligomers. The range of species (cyclic oligomers from the trimer to the octamer and linear oligomers in the range n = 7 to >20), were detected for all four rubbers.

Overall the levels of siloxane oligomers and other species (for example platinum and formaldehyde) that were found to migrate from the rubbers into food simulants were very low or below the detection limit of the experimental method used. Examples of the data obtained on the rubbers are: <0.02 mg/dm² (platinum), <0.5 mg/kg (formaldehyde), and the total amount of siloxane oligomers varied from 0.1 mg/dm² for the distilled water simulant to 58 mg/dm² for 95% ethanol.

The values obtained for the silicone oligomers can be compared to the migration limit of 10 mg/dm² stated in the Council of Europe Silicone Resolution. However, it is important to state that testing using 95% ethanol is relatively severe as silicone rubbers are known to have a high affinity for hydrophobic (i.e., fatty) media. Hence care should be taken in choosing silicone rubber for a product that would come into contact with food with a high fat content. Also, GPC analysis of the oligomers that had migrated into the 95% ethanol showed them to have an average MW of ~1,500 Daltons; a significant proportion therefore being above 1000 and so not regarded as being significant in terms of absorption in the gastrointestinal tract.

Work with the silicone rubbers using the food products showed that the total amount of silicone oligomers varied from 0.05 mg/dm² (carbonated water) to 0.56 mg/dm² (olive oil), with the concentration of the most abundant individual oligomer species (the cyclic hexamer) varying from 0.009 mg/dm² (carbonated water) to 0.03 mg/dm² (olive oil)

5.4.4.3 Data Obtained for Silicone Resins

Lower overall migration values were obtained with the silicone resins (than for silicone rubbers). This is explained by their difference in physical form, the rubbers having a lightly crosslinked mobile matrix which enables the absorption of simulant/food and the diffusion and migration of oligomers from the matrix, whereas the resins have a highly crosslinked matrix. For example, the overall migration results for the loaf pan product were 0.14 mg/dm² (10% ethanol) and 0.68 mg/dm² (95% ethanol), and the swiss roll tray gave 1.54 mg/dm² with 95% ethanol. Very small amounts of formaldehyde were detected, the highest result being 0.013 mg/dm² (swiss roll tray tested with distilled water).

With food products, no silicone oligomers were detected above the blank value (~2 µg/dm2) when testing the loaf pan with carbonated water, white wine or olive oil. Also, no oligomers were detected in carbonated water and white wine that had contacted the swiss roll tray, but 13.2 μg/dm² of oligomers in total (5.5 μg/dm² of the hexamer) were detected in olive oil.

5.4.4.4 Data Obtained for Silicone Fluids

Silicone fluids are completely different to silicone rubbers and resins due to their liquid form. Higher levels of oligomer migration have been found – particularly for the 95% ethanol fatty food simulant. Results varied from 0.09 mg/g silicone fluid used (distilled water) to 6.9 mg/g silicone fluid used (95% ethanol). As expected, the viscosity of the silicone fluid also had a significant influence on the results with the expected trend of migration increasing with reducing viscosity.

Partitioning of all three silicone fluids by the fatty food simulant was seen, with the low MW component in the original fluid being selectively extracted. This low MW component in the 95% ethanol was shown by GPC to have an average MW range around 1,500 Daltons, with the MW of the other component being around 30,000 (depending on the viscosity of the original fluid). In comparison, the oligomer fraction that migrated into distilled water was found to have average MW ranging from 17,000 to 27,000 Daltons.

Contact conditions (time and proportions of fluid to simulant) are important. The ratio of food simulant to silicone fluid had a much greater effect than the contact time. When this type of work is carried out on a specific silicone fluid product, careful consideration needs to be given to these experimental parameters. They should reflect, as closely as possible, the exact contact conditions that this product experiences in service.

Migration testing using orange juice, white wine and olive oil was carried out using the same type of contact experiment used with the simulants. The results obtained indicated that migration had occurred in both directions, i.e., components from the food had migrated into the fluid and *vice versa*. Analysis of the food products by GPC for oligomers was attempted (the work with simulants having shown that the migrants were above the MW limit for GC-MS) but, in those cases where a detected response could be obtained, components from the food were found to interfere. However, for olive oil, a peak at a MW which corresponded to the low MW peak found in 95% ethanol data was apparent. Overall, this approach showed promise for understanding the migration behaviour of the fluids, but it is apparent that further development of the method is still required.

6 Migration Mechanisms, Potential Migrants and Published Migration Data

6.1 Possible Migration Mechanisms for Chemical Species from Silicone Products

6.1.1 Migration to Air (Volatilisation)

The earliest approach (and simplest analytically) to the characterisation of migratable species from silicones was via the volatiles generated on heating. Olsen's research group in Copenhagen analysed the volatile mix from a peroxide-cure silicone rubber [41]. The curative was dicumyl peroxide and the volatile mix analysed was that collected throughout a 170 °C/20 minute cure. The species identified by GC-MS included:

- Breakdown products of dicumyl peroxide (α,α-dimethylbenzyl alcohol, acetophenone, methyl cumyl ether),

- Cyclic polydimethylsiloxanes over the range D_4 to D_{16},

- Vinyl substituted cyclic polymethylsiloxanes over the range V_nD_4 to V_nD_7, where n is 1 or 2.

A similar approach has been used by Chikishev and co-workers to evaluate the volatiles from curing of silicone rubber during both cure (120 °C/10 min), and post-cure (200 °C/6 h) using 2,4-dichlorobenzoyl peroxide as the curative [42, 43]. More than 30 chromatographic peaks attributable to organosilicon species were seen, the quantities increasing with increasing temperature of the rubber. The major species and trends identified by analysis (GC-MS) included:

- Dichlorobenzene as the principal volatile curative breakdown product - its yield increasing with extent of post-cure (reaching 20.5% of total volatile yield),

- Cyclic PDMS with D_4 and D_5 the most abundant throughout the cure and post-cure, but D_6 and D_7 increasing during post-cure,

- Vinyl substituted cyclic polymethylsiloxanes of the type VD_n with VD_2 the smallest, and

- Traces of linear dimethylsiloxanes with 2 to 7 silicon atoms.

No mention was made of any tetrachlorobiphenyl in these volatiles (see **Table 40** and surrounding discussion), possibly as a result of its limited volatility at these temperatures. Nevertheless, the yield of 2,4-dichlorobenzene was comparable with the most abundant of the cyclic siloxane oligomers at the end of the post-cure.

Chikishev and Semikin have extended the work to examine the products of migration from a silicone in an artificial breast implant [44]. The various materials analysed in this work were subjected to fractionation process described as 'gas extraction'. In this, the sample concerned was maintained at 150 °C for sequential 60 minute periods. Each of the volatile fractions obtained was then analysed by GC-MS. The components identified were all cyclic silicones of the form V_nD_m where n is 0-3 and m is 3-33.

For a silicone polymer, the first four fractions were found to contain:

- 0 - 60 min, mainly cyclic siloxanes from D_3 to D_7,

- 60 - 120 min, mainly cyclic siloxanes from D_4 to D_{10},

- 120 - 180 min, mainly cyclic siloxanes from D_5 to D_{11},

- 180 - 240 min, mainly high boiling cyclics from D_{11} to D_{33}.

6.1.1.1 Potential for Chemical Change Upon Thermal Volatilisation

A characteristic feature of the volatiles released from heated silicone is the predominance of cyclic structures. Whilst this may reflect the composition of the low MW components of the rubber, the possibility of chemical change cannot be discounted. For example, thermal depolymerisation is a route to low MW cyclic polysiloxanes and is known to be amenable to catalysis. Evidence for significant Si–O cleavage at 250 °C or above is taken from the chemical welding of heated silicone surfaces when pressed together [45]. Within the bulk of the material, and at contacting surfaces, the onset of this reaction can be regarded as a process of equilibration (i.e., Si–O/Si–O interchange). However, at exposed surfaces, this same interchange can become a process of distillation as the equilibrium is disturbed by the loss of the more volatile products.

Given that the generation of low MW material from mid-chain sections will require Si–O/Si–O interchange reactions at two separate points, the fragment released is inevitably cyclic. Only fragments released from the chain ends will be linear. On this basis, the volatile material generated by thermal or catalysed Si–O/Si–O interchange will be predominantly cyclic. This hypothesis is supported by the results of one recent study which found that 80% of the mass of a breast-implant silicone gel can be recovered as cyclic siloxanes (D_3 to D_8) when subjected to vacuum distillation at 150 – 180 °C over 24 hours [46].

6.1.1.2 Potential for Change on Ionisation

This is particularly relevant to the analysis of oligomeric siloxanes by mass spectral analysis – one of the principal techniques used in food migration work (see Section 4.1.3).

With matrix-assisted laser desorption ionisation (MALDI) techniques, a low energy ionisation is accomplished by cation attachment, so that minimal fragmentation of the organic species occurs. In the work of the Queensland group, a sodium-salt matrix was used to generate positive ions from oligomeric siloxanes at MW 23 units above the siloxane component (i.e., siloxane plus Na^+) [47, 48]. The minimal fragmentation allows for confident assignment of silicone MW.

However, the situation is different with electron impact (EI) ionisation, where a sufficient excess energy is available to cause substantial fragmentation. As silicon is more electropositive than carbon, siliconium ions would be expected, and this is borne out by one study of the EI fragmentation of cyclic polydimethylsiloxane oligomers [49]. This study found two principal modes of fragmentation:

Methyl loss $\equiv Si-Me \rightarrow \equiv Si^+ + Me\cdot$

Methyl shift (intermolecular) $(SiMe_2O)_n \rightarrow Me_3Si-(O-SiMe_2)_x{}^+$

These give rise to two different fragmentation sequences:

(i) Methyl loss from $(SiMe_2O)_n$

Gives *m/z* (n): 133(2), 207(3), 281(4), 355(5), 429(6), 503(7), 577(8), and so on.

(ii) Methyl shift to form $Me_3Si-(O-SiMe_2)_x{}^+$

Gives *m/z* (x): 73(0), 147(1), 221(2), 295(3), 369(4), 443(5), 517(6), and so on.

For this work, the spectra were obtained at relatively low ionisation energies (18 eV) and the lower *m/z* range of the cyclic siloxane spectra were characterised by strong *m/z* 73, 147 and 207 peaks. The *m/z* 207 peak corresponds to the cyclic trimer with one methyl missing (i.e., the pentamethylcyclotrisiloxane ion): it must be assumed therefore that the cyclic trimeric siloxane is particularly stable under these ionisation conditions.

Higher ionisation energies (e.g., 70 eV) are more commonplace in analytical mass spectrometry, and comparison with the reference spectra in the *Eight Peak Index* [50] reveals a less abundant *m/z* 207 ion in most cyclic oligomer spectra, except for that of hexamethylcyclotrisiloxane where it is the only abundant peak. In these cases the lower *m/z* range of the EI spectra are characterised by a dominant *m/z* 73 peak (Me shift fragment).

There is no recognisable molecular ion in the spectra of dimethysiloxane cyclic oligomers: the highest *m/z* value discernible is the M-15 (Me loss) peak.

If the same fragmentation applies to the methyl-terminated linear oligomers, then either Me loss or Me shift gives rise to the same *m/z* values, for example:

Me loss/Me shift: 73, 147, 221, 295, 369, 443, 517, 591, 665, 739, 813, 887, and so on.

It is perhaps not surprising therefore, that the low *m/z* range of the relevant reference spectra in the RSC's *Eight Peak Index of Mass Spectra* are characterised by the fragments at *m/z*: 73, 147 and 221.

Also, as with the cyclic oligomers, the spectra of these linear oligomers show no recognisable molecular ions: the highest *m/z* value discernible is the M-15 (Me loss) peak. Perhaps reflecting the stability of the cyclic siloxanes, there is evidence of some cyclisation in the fragmentation of the linear oligomers. For example both tetradecamethylhexasiloxane and hexadecamethylhepta-siloxane show significant M-103 peaks in their EI mass spectra. This appears to suggest a loss of $SiMe_4$ and Me to give the corresponding cyclic oligomer with one methyl missing, for example:

Tetradecamethylhexasiloxane → nonamethylcyclopentasilixane ion

Hexadecamethylheptasiloxane → undecamethylcyclohexasiloxane ion

Once the cyclic ring is formed, depolymerisation follows as for the cyclic oligomers, and for both the cyclic and linear oligomers there are two comparable sets of *m/z* values. For the methyl-terminated linear polydimethylsiloxanes, these are:

Me loss/Me shift: 73, 147, 221, 295, 369, 443, 517, 591, 665, 739, 813, 887, and so on.

Cyclisation: 133, 207, 281, 355, 429, 503, 577, 651, 725, 799, 873, and so on.

The differences between the spectra of the linear and cyclic oligomers are in matters of degree. For example, whilst *m/z* 73 is dominant for both linear and cyclic oligomers, the 147, 221, 295 series is more evident in the spectra of the linear oligomers than in those of the cyclics.

Another difference relates to the highest *m/z* values in these spectra. These correspond to methyl loss from the respective molecular ions (M-15 peaks). Thus, for these two types of polydimethysiloxane, the highest *m/z* values in the EI spectra would be expected at:

(i) *m/z* values in the series (74.04n – 15.02) for the cyclic polymers,

(ii) *m/z* values in the series (74.04n – 1.01) for the methyl-terminated linear polymers,

6.1.2 Migration into Fluids

Against a background of possible thermally-induced changes, it is perhaps not surprising that the composition of the migratable fraction from silicones is different when (significant) heating is not involved. Such is the case when various leachates from a breast implant were cleaned up for analysis using solvent extraction as distinct from the thermal fractionation 'gas extraction' described above [46].

In this case three leaching media were used, and the solvent extraction of the leachate was performed with an ethyl acetate/methanol (2:1) mix. The three extraction media were:

- Soy oil as an example of a lipid-rich medium,

- An aqueous culture medium [(modified Dulbecco's medium (DMEM)], and

- An emulsion consisting of DMEM plus 10% soy oil.

The findings can be summarised as follows:

- Migration levels were greatest when the surrounding medium was for the lipid-rich,

- The predominant materials released were the cyclic siloxanes, D_4 to D_8,

- Smaller amounts of higher MW material (up to 20 silicons) were obtained as both cyclic and linear polymers, and

- Traces of platinum also accompanied the migrating siloxane.

Both linear and cyclic siloxanes were found in extraction studies carried out by a research group at Queensland University in Australia [47, 48]. Slices of cured silicone rubber were subjected to a 6 hour Soxhlet extraction with chloroform. Four different rubbers were examined, and the extracts were evaporated onto a matrix for MALDI-TOF mass spectral analysis. This low-energy ionisation technique (see previous section) generates minimal fragmentation, and proved powerful enough to distinguish three different types of linear polymer:

$$Me-(SiMe_2-O)_n-SiMe_3$$

$$Me-(SiMe_2-O)_n-SiMe_2-OH$$

$$HO-(SiMe_2-O)_n-SiMe_2-OH$$

The ratios of cyclic to linear siloxanes were also determined for all the extracts, this ranging from 49 to 100% for the first extractions on each of the four polymers. The polymer delivering 100% of cyclic material was known to be 100% cyclic itself. However, of the four polymers, this gave the lowest level of extractable material. The cyclic oligomers extracted covered the range D_{14} to D_{32}. The polymer giving the highest proportion of linear siloxanes, gave the highest proportion of silanol-terminated species. The MW range of the extracts was similar for all four polymers.

Extractions with *iso*-octane, ethanol, ethanol/water, ethyl acetate and olive oil figured in studies on siloxane migration performed by the Fraunhofer Institute of Food Technology and Packaging [51]. Their studies on the three solvents (*iso*-octane, ethanol and ethyl acetate) found that the amounts extracted depended more on the thickness of the sample, than on the polarity of the solvent. For example 24 h/40 °C extractions provided yields of about 40 mg/dm^3 for silicones with a thickness of < 100 µm and one or two orders of magnitude higher for samples of around 2 mm thickness. Migration into ethanol/water mixtures decreased markedly with increasing water content: there was practically no detectable migration when the ethanol content dropped below 70%.

Migration into hot olive oil gave lower yields than for extraction into solvents. **Table 38** gives comparative data from the Fraunhofer study.

The extracted material was characterised by SFC with both flame ionisation detection and MS detection. The extracted components were found to be oligomeric siloxanes (of up to 20 $SiMe_2O$ units) by comparison with the retention times for a reference polysiloxane standard. SFC/MS analysis of this standard gave a homologous series of methyl-terminated (MD_nM) linear siloxanes.

A similar composition of siloxane oligomers was disclosed in supplier data on extracted material from a silicone elastomer [52]. Whilst a small proportion of the lower cyclics (i.e., D_4 to D_8) was found, the majority of the extract was composed of methyl-terminated linear polymers covering the range MD_3M to $MD_{21}M$ (see **Table 1**) with the most abundant components lying in the lower half of this range.

Table 38. Transfer of material from thick sample silicones by extraction by ethyl acetate and migration into olive oil [51]		
Sample	Extraction/migration yield (mg/dm^2)	
	EtOAc (24 h/40°C)	Olive Oil (1 h/121 °C)
HTV	490	29.8
RTV-2[a]	5200	230
LSR	480	53.0
Notes: [a] for moulds; LSR: Liquid Silicone Rubber - HTV type		

Migration into aqueous simulants has been evaluated for 2,4-dichlorobenzoic acid from cured silicone rubbers. This work, performed in-house at Rapra formed part of a client-specific programme [53]. The results, for migration into 10% and 90% aqueous ethanol, gave results in the range 5–65 mg/kg depending on the rubber type, and the temperature and duration of contact. The level of cure was thought to play a part here, as a higher degree of cure would be expected to generate higher levels of curative breakdown products.

6.1.3 Migration into Foodstuffs

Evidence of silicone contamination in foodstuffs has been found by ^{1}H-NMR analysis of wine and edible oil extracts [54]. This novel application of NMR spectroscopy was found to give sensitivity down to below 0.01 ppm (10 μg/kg) when using high field instrumentation and hexamethylsiloxane (HMDS) as internal standard. HMDS was chosen over tetramethylsilane (TMS) for the quantification owing to the unsatisfactorily high volatility of TMS. The difference in chemical shifts between PDMS and the HMDS was 0.8 ppm in frequency terms. Carbon tetrachloride was the extraction medium: for oil samples, a 90 min alkaline hydrolysis preceded the extraction. Up to 0.36 mg/kg of PDMS was found from the wine and up to 12 mg/kg from the edible oil. The source of the contamination was thought to be anti-foaming agents.

6.2 Potential Migrants from Silicone Products

6.2.1 Summary of Potential Migrants

The various types of silicone product (rubbers, fluids, resins and so on) that can be used in food contact situations are described in Section 2.3. For each material type,

Table 39. Potential migrants for various types of silicone product	
Silicone fluids	Silicone oligomers* and low MW siloxanes (e.g., trimethyl silanol)
Silicone rubbers	Silicone oligomers* and low MW siloxanes (e.g., trimethyl silanol) Breakdown products of peroxide curatives Platinum or tin catalysts Oxidation products (e.g., formaldehyde)
Silicone resins	Silicone oligomers* and low MW siloxanes (e.g., trimethyl silanol) Oxidation products (e.g., formaldehyde) Curing catalysts
Silicone greases, pastes and surfactants	Silicone oligomers* and low MW siloxanes (e.g., trimethyl silanol)
Up to 1000 Daltons (see Section 6.2.2.1)	

a description is given of its composition and the types of additives that can be used in it. The principal potential migrants for each major product grouping are shown in **Table 39**.

6.2.2 Specific Potential Migrants

6.2.2.1 Silicone Oligomers

Given that silicone products are relatively simple materials containing a limited number of additives and modifiers, it is this class of species that represents the source of the most important potential migrants. The types of low MW species likely to be found in silicone products are:

- TMS-terminated PDMS, MD_nM

- Cyclic PDMS, D_n

- Silanol-terminated linear PDMS, HOD_nH

- Trimethylsiloxy/silanol-terminated PDMS, MD_nH

- By-products of peroxide curatives in silicone rubbers

Each of the PDMS will provide a homologous series of components: each series distinguishable by characteristic MW. These can be derived for the various structural elements. Their precise MW can be calculated from the respective isotopic atomic weights, for example: ^1H, 1.008; ^{12}C, 12.000; ^{16}O, 15.995 and ^{28}Si, 27.997. It should be noted that other significant isotopes include: ^{13}C (1.1%), ^{18}O (0.2%), ^{29}Si (4.7%) and ^{30}Si (3.1%).

By reference to the most abundant isotopes, hexamethyldisiloxane (MM – i.e., two M units as shown in **Table 1**) has its principal MW at 162.09, whilst the basic unit D has a MW of 74.04.

Hence, the series MD_nM will have MW given by:

162.09 + 74.04n, where n is zero or an integer.

Arithmetically this rearranges to:

14.01 + 74.04x, where x = n + 2, (i.e., x is 2 or a higher integer)

Thus, the lower members of this series will have the MW – 162.1, 236.1, 310.2, 384.2, 458.25, 532.3, 606.3, 680.4, 754.4, 828.45, and so on.

The series D_n will have MW given by:

74.04n, where n is an integer (3 or above)

Thus, the lower members of this series will have the MW – 222.1, 296.2, 370.2, 444.2, 518.3, 592.3, 666.4, 740.4, 814.4, 888.5, and so on.

The series HOD_nH will have MW given by:

18.01 + 74.04n, where n is an integer

From the previous equations and the chemistry associated with the manufacture of silicone polymers (see Sections 2.1 and 2.2) it is possible to summarise the series of oligomers that can be present in the various silicone products. For example, for the PDMS, the low MW components would be expected to form a homologous series:

(i)   Cyclic oligomers: MW = 74n

(ii) Methyl only terminated linear oligomers: MW = 74.04n + 14.01

(iii) Silanol only terminated linear oligomers: MW = 74.04n + 18.01

(iv) Silanol/methyl* terminated linear oligomers: MW = 74.04n + 16.03
     (*One of each end group type)

where: n = number of repeat units in the molecule

In addition, if there are any vinyl groups (C_2H_3) in the polymer backbone, which is often the case with silicone rubbers, the MW of the oligomers will be increased by 12.00 for each vinyl unit.

If there are any phenyl groups (C_6H_5) in the polymer backbone, again a possibility with silicone rubbers, the MW of the oligomers will be increased by 62.02 for each unit.

Silicone oligomers are present in all of the silicone products (rubbers, fluids, pastes and resins) that come into contact with foodstuffs. Although these oligomers can have relatively high MW, it is widely accepted that only those having a MW up to 1000 Daltons are of interest in food migration studies.

6.2.2.2 Cure System Species in Silicone Rubbers

Peroxides are used extensively to cure silicone rubbers. These peroxides react during the cure (vulcanisation) process and leave breakdown products within the rubbers. The influence that these breakdown products have on the final product is greatly reduced by the fact that it is common practice to give silicone rubbers a post-cure in an oven after the initial vulcanisation step. This post-cure is important technologically as it improves the cure state (which improves physical properties) and in removing the breakdown products it makes the resulting product less susceptible to hydrolysis in service.

It is therefore the case that the concentration of peroxide breakdown products remaining in silicone rubbers is very low and in fact it can be difficult to detect them using conventional analysis techniques.

Peroxide breakdown products have been described in studies of polymerisation [55], vulcanisation fume [56] and in Rapra's own research programme [57]. On this basis, the likely breakdown products have either been disclosed or can be anticipated for all the above peroxides. The breakdown products are summarised in **Tables 40-44**.

Table 40. Potential breakdown products from 2,4-dichlorobenzoyl peroxide		
Species	Major/Minor	CAS number
2,4-Dichlorobenzoic acid	Major	50-80-0
2,4-Dichlorophenyl 2,4′-dichlorobenzoate	Major	
1,3-Dichlorobenzene	Major	541-73-1
2,4,2′,4′-Tetrachlorobiphenyl[a]	Major	
[a] *The potential to generate a tetrachlorobiphenyl carries a note of warning in the current BRMA Code of Practice 'Toxicity and Safe Handling of Rubber Chemicals' [58]*		

Table 41. Potential breakdown products from dibenzoyl peroxide		
Species	Major/Minor	CAS number
Benzoic acid	Major	65-85-0
Phenyl benzoate	Major	93-99-0
Benzene	Major	71-43-2
Biphenyl	Major	95-52-4

Table 42. Potential breakdown products from dicumyl peroxide		
Species	Major/Minor	CAS
α,α-Dimethylbenzyl alcohol (alternative name: 2-phenyl-2-propanol)	Major	617-94-7
α-Methylstyrene	Intermediate	98-83-9
Isopropylbenzene (cumene)	Intermediate	98-82-8
Acetophenone (acetylbenzene)	Major	98-86-2
Methyl cumyl ether	Minor	-
Dicumyl ether	Minor	-

Table 43. Some potential breakdown products from 2,5-bis(*tert*.butylperoxy)-2,5-dimethylhexane		
Species	Major/Minor	CAS number
Tert.butanol (2-methyl-2-propanol)	Major	75-65-0
Tert.butanol (2-methyl-2-propanol)	Major	75-65-0
Isobutene (2-methylpropene)	Intermediate	115-11-7
Acetone	Intermediate	67-64-1
Methyl isopropyl ether	Uncertain	598-53-8
2,5-Dihydroxy-2,5-dimethylhexane	Major	-
Hexane-2,5-dione	Intermediate	-
2-Hydroxy-2,5-dimethylhex-5-ene	Intermediate	-
2-Hydroxy-2-methylhexa-5-one	Intermediate	-
2-Methylhexa-1-ene-5-one	Intermediate	-

Table 44. Potential breakdown products from di-*t*-butyl peroxide		
Species	Major/Minor	CAS number
Tert.butanol (2-methyl-2-propanol)	Major	75-65-0
Isobutene(2-methylpropene)	Intermediate	15-11-7
Isobutane(2-methylpropane)	Minor	72-28-5
Acetone	Major	67-64-1
Methyl isopropyl ether	Uncertain	598-53-8
Methyl *tert*.butyl ether	Minor	1634-04-4
Di-*tert*.butyl ether	Minor	

Given the importance of 2,4-dichlorobenzoyl peroxide in silicone gum vulcanisation (indeed it is the only successful peroxide curative for hot air vulcanisation of extruded profiles), the manner of polychlorinated biphenyl (PCB) formation has been described by the manufacturers [59]. This work showed that the yields of both 2,4-dichlorobenzene and 2,4,2′,4′-tetrachlorobiphenyl increased with temperature over the range 100 to 325 °C. Both are products of the 2,4-dichlorophenyl radical which is produced from the initially acylperoxy radical formed by the elimination of CO_2, for example:

$$XC(=O)OOC(=O)X \rightarrow 2XC(=O)O\bullet$$
$$XC(=O)O\Sigma \rightarrow X\bullet + CO_2$$

where: $X = C_6H_3Cl_2-$

Clearly increasing the temperature promotes this elimination.

The potential for this peroxide to produce a PCB compound has encouraged the use of other peroxides (e.g., dicumyl peroxide), or platinum-based cure systems, for food contact rubbers.

The other major cure system that is used in food contact silicone rubbers uses metal catalysts. The majority of these are based on platinum, with the use of compounds such as chloroplatinic acid. This is an area of active research and the precise platinum compound used by particular companies is a closely guarded secret. Fortunately, the metal is easily detectable in food and food simulants by established techniques such as ICP (see Section 4.1.2) and limits for platinum itself are given in food contact regulations, e.g., the German *BfR Recommendation XV* (see Section 3.3).

6.2.2.3 Low MW Products Formed Due to Oxidation Reactions

Silicone rubbers and resins are both manufactured (processed and cured) and used in service at high temperatures. Although they are sufficiently stable to high temperatures (e.g., 200 °C) over relatively long periods of time (e.g., weeks) and retain good physical properties, chemical changes can occur within the materials if this period is extended to a number of months or years. There are two mechanisms that can generate low molecular species having the potential to migrate into food.

Main chain scission can occur leading to cyclisation and the formation of relatively low MW cyclic oligomers.

Oxidation of the alkyl groups attached to the silicone atoms can lead to the formation of oxidation products such as aldehydes (e.g., formaldehyde from methyl groups).

6.3 Published Migration Data

6.3.1 Silicone Rubber Study

A test report has been produced by the Fraunhofer Institute [60] on the migration of siloxanes from three different silicone rubbers: a high temperature curing material, a room temperature curing material, and a cure liquid silicone rubber. Five different food simulants (*iso*-octane, ethanol, ethanol-water, ethyl acetate and olive oil) were used and one of the things investigated was the degree to which the thickness of the sample affects overall migration. This was found to be more important than the polarity of the simulant, in the case of the hydrophobic solvents. As expected, the results obtained with ethanol-water mixtures showed that the amount of migrating oligomeric material reduced markedly with increasing water content, a virtually zero result being obtained above 30%. The migrants were characterised by supercritical fluid chromatography SFC using both flame ionisation and MS detection. A homologous series of methyl-terminated linear siloxane oligomers up to twenty $SiMe_2O$ units were identified.

6.3.2 Silicone Rubber Teats and Soothers

A Dutch retail survey conducted in 2003 [30] investigated the potential migrants that were present in a randomly selected sample of nineteen teats and soothers. Some of these products were made from compounds based on natural rubber (NR), in which case N-nitrosamines, N-nitrosatable substances and 2-mercaptobenzene were targeted. The majority, however, had been manufactured using silicone rubber and the only extractable species from these were found to be siloxane oligomers.

A study has also been carried out in Japan [61] on the migration of species from teats and soothers made from a number of rubbers including silicone rubber and natural rubber. The natural rubber samples gave the highest overall migration results, and also registered positive for *N*-nitrosamines. The results obtained show why silicone based teats and soothers are gaining commercial ground over more traditional materials such as NR.

6.3.3 Peroxide Breakdown Products

Peroxides are often used to cure silicone rubbers and acidic species are among the breakdown products of these compounds. A Japanese study [62] obtained data on such compounds in extracts obtained from silicone teats and jar seals using thin layer chromatography and UV absorption chromatography. The amount of 2,4-dichlorobenzoic acid in products that had not been post-cured varied from 7.7 mg/kg to 24.2 mg/kg. The lowest values were obtained using water as the extractant and the highest using *n*-pentane. Postcuring, which is usually carried out on food contact silicone products, significantly reduced the concentration of this compound.

6.3.4 Polydimethylsiloxane Oligomers

A review of the use of a number of analytical techniques (IR, GC-MS, NMR, atomic absorption spectrophotometry) to identify and quantify PDMS in a wide range of matrices (e.g., food products, pharmaceuticals and cosmetics) has been published recently [63]. This paper also considers the toxicological issues surrounding PDMS.

NMR spectroscopy has been used to quantify residual amounts of PDMS anti-foaming agents in wine and edible oil. Careful preparation and analysis techniques enabled trace levels of PDMS to be detected below the regulatory limit of 10 mg/kg food [54].

6.3.5 General Assessment of Silicone Rubbers

The food safety aspects of silicone rubbers have been reviewed by Cassidy [64]. In addition to looking at potential food migrants, the review covered a number of areas of potential hazard (e.g., biomedical contact, fire and biodurability) and concluded that silicone elastomers exhibit significantly more benign characteristics than other competing elastomers offered for the same end use applications.

7 Improving the Safety of Silicones for Food Use and Future Trends

7.1 Silicone Foams

It is possible to produce silicone sponge rubber products using chemical blowing agents (compounds that breakdown upon heating to produce low MW gaseous products), but there are either toxicity or technological/processing concerns associated with these products, and few have food use approval. This situation has been addressed by Dow Corning who have developed a technology that is capable of producing closed cell silicone sponges using water as the blowing agent [65, 66].

German-based silicone manufacturer, Wacker, has recently announced the launch of a novel silicone foam. In addition to being resistant to acids and bases, it has the ability to be stretched up to 15 times its normal size over a wide temperature range (–50 to 300 °C). Although not specifically targeted at the food contact applications market, it is suitable for use in this area [67]. Other new, food-contact silicone foam products have also been announced recently by Wacker, including one that claims to be the first bubble-free extrudable product having a hardness of 10 Shore A [68].

7.2 Antibacterial Additives and Coatings

A relatively new class of additive that has been developed for food contact rubbers is the antimicrobial agent. One company, Milliken, have introduced a family of these compounds, based on silver ion exchange resins that can be used in peroxide cured rubbers such as silicones, ethylene propylene diene monomer terpolymer and nitriles [69]. One of the prime advantages of such additives is that they control microbial growth on and within the surface of such rubbers when they are used in food production lines and this reduces the need for cleaning and part replacement. Another company which has introduced these types of additives into the market place is Sanitised AG. It has achieved US Environmental Protection Agency registration for its Sanitised PL product and has launched others, such as Sanitised Silver [70].

Another way of controlling the growth of micro-organisms is to coat the surface of a silicone product (e.g., a food-grade silicone rubber) with an anti-microbial coating. One route that has been developed involves depositing silver nanoparticles onto the surface of the rubber under formaldehyde - radiofrequency plasma conditions. The bacterial properties of the coated surfaces were investigated by exposing them to *Listeria moncytogenes*, with no bacteria being detected after exposure times of 12 to 18 hours [71].

In addition to being used for products which come in direct contact with food, antimicrobial additives are also used in indirect contact products. For example, their use in thermoplastic rollers that were an integral part of a food conveyor system had the desired affect of achieving an overall reduction in microbiological contamination, in addition to additional benefits such as improving lubricity and reducing wear [72].

7.3 Intelligent Packaging

A considerable amount of work is going into the development of 'Intelligent' or 'Active' food packaging materials. Work by the Landec Corporation [73] has shown that microporous membranes coated with certain silicone materials can produce packaging materials with very high permeabilities, which vary with temperature, and which can produce selective carbon/dioxide oxygen permeability ratios. Packages can be tailor made therefore to suite the specific requirements of different foodstuffs.

7.4 Barrier Coatings

In contrast to the coatings that are highly permeable (see previously) it can be important in food packaging to have available, coatings that provide a barrier to permeation. The Fraunhofer Institute [74] have developed siloxane-type barrier coating resins that are applied to a substrate by means of spraying or roller coating and then cured by the application of moderate heat (e.g., 100 °C) or UV radiation.

7.5 Non-stick Additives

The low coefficient friction of silicone products has made them ideal for a number of release and non-stick applications (e.g., silicone resin coated cookware). GE Silicones [75] have produced a spherical, fine particle silicone resin additive (trade name: Tospearl) that is intended to reduce adhesion between the surfaces of the product itself and between the surface of the product and other substrates. This additive, which is available in a number of different grades, can be used in a wide range of polymer products, including rubbers, plastics, inks and paints. Being crosslinked it does not migrate like silicone fluid type additives, and it has approval for use in food packaging applications.

Another silicone additive designed to modify the surface characteristics of polymers has been developed by Dow Corning [76]. It is available as a pelletised masterbatch, comprising of 50% silicone polymer within a polypropylene carrier. It can be added to polypropylene or similar thermoplastics and has been approved for food contact applications.

7.6 Nanoparticulate Silicones

Nanoparticle PDMS resins have been developed that can be used as surface coatings to reduce surface adhesion by as much as a factor of eight when compared to conventional silicone treated surfaces. These ultra-low, micro-structured adhesive coatings can be applied to polymers such as polyethylene, and are thought to have potential in both the food packaging and food processing sectors [40].

7.7 Inks and Varnishes

The introduction of food contact regulations such as *Council of Europe Resolution on Packaging Inks* has caused some concern amongst the manufacturers and users of inks and varnishes [77]. Manufacturers such as Dow Corning [78] have addressed these concerns and shown how a range of water-borne silicone copolymer products can be produced that combine both good processing and application properties with food contact approval.

7.8 Radiation-cured Release Coatings

This type of release coating (cure using either radical or cationic mechanisms) is gaining popularity over thermally cured release coatings due to the increasing preference of industry for thermally sensitive polymer substrate/pressure sensitive adhesive combinations. Fundamental research is underway to understand how the radiation cure chemistry affects the properties and performance of these types of coatings [79].

8 Conclusion

This review has given an overview of the extensive range of silicone products that are used in food contact situations, together with an introduction to the chemical technology that is associated with their manufacture, and an overview of the food contact legislation that is associated with these types of materials. It has also provided a summary of the analytical techniques and approaches that are used to assess the food safety of silicone products, the potential migrants that can be present in these materials, and the migration data that is in the public domain. It also contains a summary of the recent developments that have been associated with silicone-based food contact materials.

One of their most important properties, very good heat resistance, means that silicones have a simpler composition than most other polymer systems because they do not require stabilisers, and their cure systems, when required, are also simpler. The result of this relative simplicity is that they contain a smaller range of potential migrants and, because crosslinked products are often post-cured to ensure a good cure state, with low levels of volatiles, the products are relatively 'clean'. Despite this relative 'cleanness', in common with the majority of food contact materials, silicone food contact legislation has been an active area recently, with the adoption of a *CoE Resolution on Silicone Rubbers* in 2004.

The principal low MW migrants that are present in silicones are siloxane oligomers. These can present a challenge for the analyst, usually having similar mass spectra to one another, so it can be difficult to determine MW values, and no UV absorbing group. For their determination, and that of other specific migrants from silicones (e.g., peroxide residues), the commercial proliferation of LC-MS instruments, with their enhanced capability compared to HPLC systems, and the new generation of GC-MS instruments (e.g., GCxGC-ToFMS) are of great benefit to analysts who are carrying out food migration and other tasks (e.g., reverse engineering and failure diagnosis).

Their range and versatility, coupled with important attributes (e.g., a high level of thermal stability, good low temperature flexibility, and low surface free energy) will continue to ensure that silicones products are increasingly used for food contact applications that require properties such as high purity, good heat resistance, and low levels of surface adhesion.

Acknowledgements

The author would like to acknowledge the contribution made to this report by Dr Bryan Willoughby. This arose out of the work that he was commissioned to do by Rapra Technology for the FSA Silicones project (Contract number A03046). In particular Dr Willoughby's work has been used in the writing of Chapter 2 and parts of Chapter 6.

Thanks are also expressed to Wacker Chemicals Ltd, Silicones Division for the provision of representative food grade silicone rubbers, fluids and resins for use in the FSA project and for the help and assistance that they gave during the course of the project. Particular thanks are due to Wacker UK Sales Manager, Ashish Sachdeva and his predecessor, Malcolm Harrison, for their interest and technical support.

Finally, the author would also like to thank the UK Food Standards Agency for the funding to carry out the Silicones project.

Structural Assignments for Silicone Polymers and Oligomers

Structure	Symbol
R_3Si-O	M
$-O-SiR_2-O-$	D
$RSi(-O-)_3$	T
$Si(-O-)_4$	Q

References

1. *Chemistry and Technology of Silicones*, Ed., W. Noll, Academic Press, New York, NY, USA, 1968, p.1.

2. *Chemistry and Technology of Silicones*, Ed., W. Noll, Academic Press, New York, NY, USA, 1968, p.6.

3. A.F.M. Barton, *CRC Handbook of Solubility Parameters and Other Cohesion Parameters*, CRC Press, Boca Raton, FL, USA, 1968, p.406.

4. *Chemistry and Technology of Silicones*, Ed., W. Noll, Academic Press, New York, NY, USA, 1968, p.5.

5. K.J. Saunders, *Organic Polymer Chemistry*, Chapman and Hall, London, UK, 1973, p.15.

6. *Chemistry and Technology of Silicones*, Ed., W. Noll, Academic Press, New York, NY, USA, 1968, p.2.

7. J.A.C. Watt, *Chemistry in Britain*, 1970, **6**, 519.

8. C. Eaborn, *Organosilicon Compounds*, Butterworths, London, UK, 1960, p.246.

9. R. Meals, *Silicon Compounds (Silicones) in Kirk-Othmer Encyclopedia of Chemical Technology*, Volume 18, Ed., A. Standen, Interscience, New York, NY, USA, 1963.

10. *Chemistry and Technology of Silicones*, Ed., W. Noll, Academic Press, New York, NY, USA, 1968, p.9.

11. W.J. Bobear in *Rubber Technology*, Ed., M. Morton, Van Nostrand Reinhold, New York, NY, USA, 1973, p.15.

12. M. Schaetz, *Plasty a Kaucuk*, 1991, **28**, 1, 5.

13. M. Schaetz, *International Polymer Science and Technology*, 1992, **19**, 2, 33.

14. P. Jerschow, *Silicone Elastomers*, Rapra Technology Ltd, Shawbury, UK. Rapra Review Report, 2001, **137**, 12, 5.

15. *Tin Chemicals for Today's World: The Formula for Success*, ITRI Publication, The International Tin Research Institute, Uxbridge, UK, 1997, p.681.

16. *Proctor & Hughes' Chemical Hazards of the Workplace*, Eds., G. J. Hathaway, N. H. Proctor and J.P. Hughes, 4th Edition, Van Nostand Reinhold, New York, NY, USA, 1996.

17. J.K. Gillham, in Proceedings of Rapra Technology Ltd Conference – *Flow and Cure of Polymers-Measurement and Control*, Shawbury, UK, 1990, Paper No. 6a.

18. *Chemistry and Technology of Silicones*, Ed., W. Noll, Academic Press, New York, NY, USA, 1968, Chapter 4.

19. K.E Polmanteer, *Rubber Chemistry and Technology*, 1981, **54**, 5, 1051.

20. J. Stein, L.N. Lewis, Y. Gao and R.A Scott, *In Situ Determination of the Active Catalyst in Hydrosilation Reactions using Highly Reactive Pt(0) Catalyst Precursors*, GE R&D Technical Report 98CRD115, General Electric Company, USA, 1998.

21. F.A. Cotton and G. Wilkinson, *Advanced Inorganic Chemistry*, 5th Edition, Wiley, New York, NY, USA, 1988, p.917 and p.1255.

22. L.N. Lewis, J. Stein, R.E. Colborne, Y. Gao and J. Dong, *The Chemistry of Fumarate and Maleate Inhibitors with Platinum Hydrosilation Catalysts*, GE R&D Technical Report 96CRD030, The General Electric Company, USA, 1996.

23. A. Karlsson, S.K. Singh and A-C Albertsson, *Journal of Applied Polymer Science*, 2002, **84**, 12, 2254.

24. S. Smith in Preparation, *Properties and Industrial Applications of Organofluorine Compounds*, Ed., R. E. Banks, Ellis Horwood, Chichester, UK, 1982.

25. *Chemistry and Technology of Silicones*, Ed., W. Noll, Academic Press, New York, NY, USA, 1968, Chapter 10.

26. *Chemistry and Technology of Silicones*, Ed., W. Noll, Academic Press, New York, NY, USA, 1968, Chapter 8.

27. S.F. Thames in *Applied Polymer Science*, Eds., R.W. Tess and G.W. Poehlein, ACS Symposium Series No. 285, American Chemical Society, Washington, DC, USA, 1985.

28. P. Dufton, *Functional Additives for the Plastics Industry*, Rapra Technology Ltd, Shawbury, UK, 1998, 284.

29. *The Polyurethanes Book*, Ed., D Randall and S. Lee, John Wiley & Sons Ltd., Chichester, UK, 2002.

30. K. Bouma, F.M. Nab and R.C. Schothorst, *Food Additives and Contaminants*, 2003, **20**, 9, 853.

31. A.A. Zotto in *Food Additive User's Handbook*, Ed., J. Smith, Blackie, Glasgow, UK, 1991.

32. M. Suman, *British Pharmacopeia*, 2008, British Pharmacopeia Secretariat, London, UK, 2007.

33. CEN 13130-23, *Materials and Articles in Contact with Foodstuffs - Plastics Substances Subject to Limitation, Part 23: Determination of Formaldehyde and Hexamethylenetetramine in Food Simulants*, 2005.

34. *International Journal of Polymeric Materials*, 2003, **52**, 1, 1.

35. M.J. Forrest, S.R. Holding, D. Howells and M. Eardley in *Proceedings of a Rapra Conference on Silicone Elastomers*, Frankfurt, Germany, 2006, p.3.

36. CEN 13130-23, *Materials and Articles in Contact with Foodstuffs - Plastics Substances Subject to Limitation, Part 23: Determination of Formaldehyde and Hexamethylenetetramine in Food Simulants*, 2005.

37. M. Forrest, S. Holding and D. Howells in *Proceedings of a Rapra Technology Conference – High Performance and Speciality Elastomers*, Geneva, Switzerland, 2005, Paper No. 2.

38. J. Sidwell in *Proceedings of a Rapra Technology Conference – RubberChem 2002*, Munich, Germany, 2002, 16.

39. K.A. Barnes, L. Castle, A.P. Damant, W.A. Read and D.R. Speck, *Food Additives and Contaminants*, 2003, **20**, 2, 196.

40. S. Loher, WJ. Stark, T. Maienfisch, S Bokorny and W. Grimm, *Polymer Engineering and Science*, 2006, **46**, 11, 1541.

41. H. Olsen, in *Proceedings of the 8th Scandinavian Rubber Conference*, Copenhagen, Denmark, 1985, p.571.

42. Yu.G. Chikishev, V.V. Semikin, B.E. Gadas, *International Polymer Science and Technology*, 1985, **12**, 6, 34

43. Yu.G. Chikishev, V.V. Semikin, B.E. Gados, *Kauchuk i Rezina (USSR)*, 1985, **2**, 30.

44. Yu.G. Chikishev, V.V Semikin, *International Polymer Science and Technology*, 1985, **12**, 12, 14.

45. C.A.P. Leite, R.F. Soares, M. do C. Goncalves and F. Galembeck, *Polymer*, 1994, **35**, 15, 3173.

46. E.D. Lykissa, S.V. Kala, J.B. Hurley and R.M. Lebovitz, *Analytical Chemistry*, 1997, **69**, 23, 4912.

47. S. Hunt, G. Cash, H. Liu, G. George and D. Birtwistle, *Journal of Macromolecular Science A*, 2002, **39**, 9, 1007.

48. S.M. Hunt and G.A. George, *Polymer International*, 2000, **49**, 7, 633.

49. A. Ballistreri, D Garozzo and G.Monaudo, *Macromolecules*, 1984, **17**, 7, 1312.

50. *The Eight Peak Index of Mass Spectra*, 3rd Edition, Royal Society of Chemistry, Nottingham, UK, 1983.

51. O. Piringer and T. Bucherl, *Extraction and Migration Measurements of Silicone Articles and Materials Coming into Contact with Foodstuffs*, FhG Test Report, Munich, Germany, 1994.

52. Letter from Dow Corning Ltd to Rapra Technology Ltd, 1997.

53. J.A. Sidwell and A. M. Jolly, *Food Contact Elastomers Group: Studies on Dow Corning Standard Samples*, Rapra Confidential Technical Report, Rapra Technology, Shawbury, UK, 1996.

54. K. Mojsiewicz-Pienkowska, Z. Jamrogiewicz and J. Lukasiak, *Food Additives and Contaminants*, 2003, **20**, 5, 438.

55. R.W. Lenz, *Organic Chemistry of Synthetic High Polymers*, Interscience, New York, NY, USA, 1967.

56. B.G. Willoughby, *Progress of Rubber Technology*, Elsevier, Applied Science Publishers Ltd, Barking, UK, 1984, **46**, 143.

57. B. Willoughby, *Rubber Fume: Ingredient Emission Relationships*, Rapra Technology Ltd, Shawbury, UK, 1994.

58. *Toxicity and Safe Handling of Rubber Chemicals – BRMA Code of Practice*, Rapra Technology, Shawbury, UK, 1999.

59. J.D. van Drumpt in *Proceedings of the 136th ACS Rubber Division Meeting*, Detroit, MI, USA, Fall 1989, Paper No. 116.

60. O. Pringer and T. Bucherl, *Extraction and Migration Measurements of Silicone Articles and Materials Coming into Contact with Foodstuffs*, FhG Test Report, 1994.

61. K. Mizuishi, M. Takeuchi, H. Yamanobe and Y. Watanabe, *Annual Report of Tokyo Metropolitan Research Laboratory of Public Health*, 1986, **37**, 145.

62. T. Baba, K. Kusumoto and Y. Mizunoya, *Journal of the Food Hygienics Society of Japan*, 1979, **20**, 5, 332.

63. K. Mojsiewicz-Pienkowska and J. Lukasiak, *Polimery*, 2003, **48**, 6, 401.

64. S.L. Cassidy, *Progress in Rubber and Plastics Technology*, 1991, 7, 4, 308.

65. R. Romanowski, B.A. Jones and T.J. Netto in *Proceedings of the 164th ACS Rubber Division Meeting*, Cleveland, OH, USA, Fall 2003, p.66.

66. E. Gerlach and F. Giambelli in *Proceedings of a Conference – IRC 2003*, Nuremberg, Germany, 2003, p.339.

67. J. Weidinger, *Wacker World Wide*, 2006, **6**, 2, 7.

68. *Chemical Weekly*, 2005, **51**, 11, 167.

69. B. Patel, S. McDowell, R.C. Kerr and G.R. Haas in *Proceedings of the 164th ACS Rubber Division Meeting*, Cleveland, OH, USA, Fall 2003, 27.

70. *Additives for Polymers*, 2003, October, p.3.

71. H. Jiang, S. Manolache, A.C.L. Wong and F.S. Denes, *Journal of Applied Polymer Science*, 2004, **93**, 3, 1411.

72. *Plastics Engineering*, 2003, **59**, 6, 17.

73. R Clarke, *Journal of Plastic Film and Sheeting*, 2001, **17**, 1, 22.

74. *Chemical Engineering*, 1996, **103**, 8, 19.

75. F. Schlossels and M. Ohtsuki, *Addcon '96. Conference Proceedings*, Brussels, 1996, 5.5.

76. *Dow Corning MB50-001 Silicone Masterbatch – Product Information*, Dow Corning Europe, La Hulpe, Brussels, Belgium, 1995.

77. M. Forrest, *Coatings and Inks for Food Contact Materials*, Rapra Review Report No. 186, Rapra Technology, Shrewsbury, UK, 2007.

78. V. James, Polymers Paint Journal, 2006, **196**, 4504, 48.

79. G.V Gordon, P.A. Moore, P.J. Popa, J.S. Tonge and G.A. Vincent, *Adhesives Age*, 2002, **45**, 7, 24.

Abbreviations and Acronyms

ADI	Acceptable daily intake
Amu	Atomic mass unit
APCI	Atmospheric pressure chemical ionization
ASTM	American Standards Testing Materials
ATD	Automated thermal desorption
ATR	Attenuated total reflectance
BADGE	Bisphenol A diglycidyl ether
BCF	British Coatings Federation
BFDGE	Bisphenol F diglycidyl ether
BfR	Bundesinstitut fur Risikobewertung
BHT	Butylated hydroxyl toluene
bp	Boiling point
BPA	Bisphenol A
BPF	Bisphenol F
CAB	Cellulose acetate butyrate
CAS	Chemical Abstracts Service
CEN	Comité Européen de Normalisation (European Committee for Standardisation)
CEPE	The European Council of Paint, Printing Inks and Artists' Colours Industry
CITPA	International Confederation of Paper and Board Converters in Europe
CoE	Council of Europe
D4	Octamethylcyclotetrasiloxane

DCM	Dichloromethane
DEFRA	Department for Environmental, Food and Rural Affairs
DETX	2,4-Diethylthioxanthone
DICY	Dicyanidiamide
DMEM	Modified Dulbecco's medium (an aqueous culture medium)
DMPA	2,2-Dimethoxy-2-phenylacetophenone
DNQ	Detected but not quantified
DPGDA	Dipropylene glycol diacrylate
DRD	Draw redraw
DTA	Diethylene triamine
DWL	Drawn and wall ironed
EB	Electron beam
EC	European Commission
EFSA	European Food Safety Association
EFSA	European Food Safety Authority
EHDAB	2-Ethylhexyl-4-dimethylaminobenzoate
EI	Electron impact ionisation
EOA	Ethylene diamine
EOE	Easy open end
EPDM	Ethylene-propylene-diene terpolymer
EU	European Union
EuPIA	European Printing Ink Manufacturers Association
EVA	Ethylene vinyl acetate
EXAFS	Extended X-ray absorption fine structure spectroscopy
FDA	The US Food and Drug Administration
FOA	Food and Drugs Administration
FOE	4′,4′ Diamino diphenyl-methane
FSA	Food Standards Agency
FSMD	Flat sheet metal decorating

FTIR	Fourier transform infrared spectroscopy
GC	Gas chromatography
GC-MS	Gas chromatography - mass spectrometry
GC-TOFMS	Gas chromatography-time of flight mass spectroscopy
GMP	Good manufacturing practice
GPC	Gel permeation chromatography
GPTA	Propoxylated glyceryl triacrylate
GRAS	Generally recognised as safe
HMDS	Hexamethylsiloxane
HTV	High temperature vulcanisation
HPLC	High performance liquid chromatography
IARC	International Agency for Research on Cancer
ICP	Inductively coupled plasma
IR	Infrared
ITX	2-Isopropylthioxanthone
LC	Liquid chromatography
LC-MS	Liquid chromatography-mass spectroscopy
LSR or LR	Liquid silicone rubber
MAFF	Ministry of Agriculture, Fisheries and Food
MALDI-TOF	Matrix assisted laser desorption ionisation – time of flight mass spectroscopy
MDM	Octamethyltrisiloxane
MM	Hexamethyldisiloxane
Mn	Number average molecular weight
Mp	Melting point
MS	Mass spectrometry
MW	Molecular weight(s)
Mw	Weight average molecular weight
NC	Nitrocellulose
Nd	Not detected by the method

N-ETSA	*n*-Ethyl-*o/p*-toluene-sulfonamide
NMR	Nuclear magnetic resonance spectroscopy
NOGE	Novolal glycidyl ethers
NP	Nonyl Phenol
NR	Natural rubber
OML	Overall migration limit
PCB	Polychlorinated biphenyl
PDMS	Polydimethylsiloxane(s)
PE	Polyethylene
PET	Polyethylene terephthalate
PETA	Pentaerythritol tetra-acrylate
PF	Phenol-formaldehyde
Phr	Parts per hundred of rubber
PP	Polypropylene
ppb	Parts per billion
ppm	Parts per million
PT	Press twist
PTFE	Polytetrafluoroethylene
PU	Polyurethane(s)
PVAc	Polyvinyl acetate
PVB	Polyvinyl butyral(s)
PVC	Polyvinyl chloride
PVOH	Polyvinyl alcohol
QC	Quality control
QMA	Maximum permitted quantity of the substance in the finished material or article expressed as mg per 6 dm^2 of the surface in contact with foodstuffs
QSAR	Quantitative structure-activity relationship.
REACH	Registration, evaluation and authorisation of chemicals
ROPP	Roll-on pilfer-proof

RT	Retention time
RTO	Regular twist off
RTV	Room temperature vulcanising
SCF	Scientific Committee for Food (Pre EFSA)
SML	Specific migration limit(s)
TDI	Tolerable daily intake
T_g	Glass transition temperature
TIC	Total ion chromatogram
TMS	Tetramethylsilane
UV	Ultra-violet
VOC	Volatile organic compound(s)
w/b	Water based
w/w	Weight/weight
XPS	X-ray photoelectron spectroscopy

Subject Index

A

Accelerators, 40 56 60 112 114
Acrylates, 215
Acrylic rubber, 13 31 59
Active packaging, 235 236 238
Alkyd resins, 148
Amines, 116 216
Amino resins, 149
Antidegradants, 14 18 40 55 58 59 95 112 211
Antimicrobial systems, 234
Antimicrobial technologies, 235
Atmospheric chemical ionisation, 281
Atmospheric pressure chemical ionisation, 37
Atomic absorption spectrophotometry, 112 212 340
Attenuated total reflectance infrared spectroscopy, 276

B

BADGE, 219 220 222
Barrier coatings, 342
BFDGE, 220
Binder resins, 165
Biodegradability, 234
BPA, 219 220 221
BPF, 220
Bundesinstitut für risikobewertung (BfR), 22, 123
  Regulations, 23 24 25 28 44 203
  Tests, 46
Butyl rubber, 13 32 59 126

C

Cellulosics, 153
Coatings, 164
Coatings, safety of, 232 233
Colorants, 168

I

Inductively coupled plasma spectrometry, 98
  Scans, 275
Inkjet, 194
Inks, application techniques of, 191
Inks constituents of, 165
Inks, safety of, 232 233
Intelligent packaging, 342
Ion mass spectroscopy, 220
Infra red spectroscopy, 340

K

Karstedtís catalyst, 265

L

Lactococcus lactis, 235
Laser marking, 235
Liquid chromatography-mass spectrometry, 36 37 40 57 58 61-63 68 70 74 75 81 84
  94 96 113 121 122 210 212 213 236 237 276 280-282 286 290 291 324
LC-MS chromatogram, 80-84 292
  Diethyl ether, 38
  Nitrile compound 351, 67
Liquid silicone cure, 262
Liquid silicone systems, one-pack, 262
Liquid silicones, 264
Lithography, 186 191 192

M

Mass spectral analysis, 329
Mass spectrometry, detection, 111
Matrix-assisted laser desorption ionisation, 329
  Time-of-flight mass spectral analysis, 332
Metal packaging, 172 173 174 175 176 179 182 186 194 200
Migration testing, 27 29 44 45 51 69-71 207 209 282 284 294 299
Monomers, 39

N

Nanotechnology, 234 236
Natural rubber, 10 11 26 52 54 55 56 59 61 107 108 135 136
Natural rubber, compounds, 42 45 46 48
Nitrile compounds, 60 62 68 351,
Nitrile rubber, 11 26 30 43 45 46 49 54 59 60 61 132 -134

CPSIA information can be obtained at www.ICGtesting.com
Printed in the USA
LVOW071046050712

288840LV00001B/5/P

9 781847 351418